Praise for *Sludge*

'This is the book about the goldfields I most wanted to read but
didn't think could be written. It's a remarkable achievement.'
—TOM GRIFFITHS, author of *The Art of Time Travel*

'If Victorians dreamed of glittering gold, what they got was
a tidal wave of sludge that covered the land like a poisonous
blanket and made the rivers run thick as gruel. Susan Lawrence
and Peter Davies vividly recreate the forgotten landscapes of
nineteenth-century Victoria, revealing how people and mining
destroyed the country that nurtured them, and how that silent
legacy is still with us today. Here is a powerful parable, a work of
brilliant rediscovery and a wake-up call for our own times.'
—GRACE KARSKENS, author of *The Colony*

'*Sludge* is a fascinating, entangled story of human
endeavour and environmental destruction. An exciting and
timely reminder that history is a dirty business, precisely
because it oozes its way into the present.'
—CLARE WRIGHT, author of *The Forgotten Rebels of Eureka*

'Sludge, slurry, slickens or porridge: call it what you will, mining
waste made a mess of Victoria's environment. In *Sludge*, Susan
Lawrence and Peter Davies carefully investigate this murky
history of greed, mismanagement, reform and forgetting. It is a
gripping account of an environmental catastrophe, and it vividly
conveys the long-term costs of short-term gains.'
—BILLY GRIFFITHS, author of *Deep Time Dreaming*

SLUDGE

SLUDGE

DISASTER ON VICTORIA'S GOLDFIELDS

SUSAN LAWRENCE & PETER DAVIES

LA TROBE
UNIVERSITY PRESS

IN CONJUNCTION WITH BLACK INC.

Published by La Trobe University Press in conjunction with Black Inc.
Level 1, 221 Drummond Street
Carlton VIC 3053, Australia
enquiries@blackincbooks.com
www.blackincbooks.com
www.latrobeuniversitypress.com.au

La Trobe University plays an integral role in Australia's public intellectual life, and is
recognised globally for its research excellence and commitment to ideas and debate.
La Trobe University Press publishes books of high intellectual quality, aimed at general
readers. Titles range across the humanities and sciences, and are written by distinguished
and innovative scholars. La Trobe University Press books are produced in conjunction
with Black Inc., an independent Australian publishing house. The members of the LTUP
Editorial Board are Vice-Chancellor's Fellows Emeritus Professor Robert Manne and
Dr Elizabeth Finkel, and Morry Schwartz and Chris Feik of Black Inc.

9781760641108 (paperback)
9781743821091 (ebook)

A catalogue record for this
book is available from the
National Library of Australia

Cover design by Regine Abos
Text design and typesetting by Akiko Chan
Cover image of gold prospectors in Queensland courtesy of KPGA/Alamy

Printed in Australia by McPherson's Printing Group.

For our parents

CONTENTS

Introduction ... 1

1. Sludge ... 9

2. Mining ... 35

3. Water ... 67

4. Fist Fights and Water Rights ... 104

5. The Sludge Question ... 136

6. Turning the Tide ... 178

7. Aftermath ... 209

Conclusion ... 248

Acknowledgements ... 255

Picture Credits ... 259

Notes ... 261

Index ... 293

INTRODUCTION

Bento Rodrigues, Brazil, 6 November 2015

WET, ORANGE MUD COVERS EVERYTHING: STREETS, houses, cars, animals, trees, fields. The violent force of a torrent of mud has overturned cars and left them hovering on top of buildings. It has torn the roofs off houses and pushed over their walls. The view of the town from helicopters flying above reveals a desolate landscape: sludge-caked animals struggle to free themselves, and rescue teams search desperately for survivors. Mud dyes the river orange for hundreds of kilometres downstream, and two weeks later it will flow out into the Atlantic in an expanding orange stain.

This devastation is the result of the catastrophic failure of a tailings dam: a vast settling pond built to store the muddy waste from Samarco's Germano iron ore mine. Late one afternoon in November 2015 the dam wall gave way. The collapse released a flood of polluted, sediment-laden water that raced down the valley below, destroying and burying everything in its path and leaving twenty-one people dead. The valley will never be the same.

Just three years later another tailings dam failed in the same part of Brazil, with more tragic human consequences. On 29 January 2019, as mine workers were sitting down to lunch in the company cafeteria, one of the dams gave way at the Córrego do Feijão iron mine near Brumadinho. Nearly twelve million cubic metres of polluted mud rushed down the Rio Paraopeba River, submerging the mine's administration area and parts of nearby communities. More than 300 people were killed, including the workers eating their lunch and two holidaymakers from Sydney. A third of the bodies still have not been recovered from the sludge.[1]

Ironically, the mine owners – which include Australia's BHP – built these tailings dams to protect the environment. Without them, waste would have poured into the river every day the mine operated. The sludge would have flowed everywhere, oozing through streets and under doors, creeping up walls and between trees, covering gardens and crops. It would still have turned the ocean orange – it just would have taken longer and occurred more gradually, over years rather than hours, and no one would have drowned.

Chewton, Victoria, Australia, 9 December 2010

Forest Creek flows placidly down a valley lush with grass and dotted with trees. Plastic guards protect seedlings planted recently as part of a program to reintroduce native species. A cycling path meanders across the low hills in the middle distance, and sheep graze in a far paddock. In the foreground two little boys play cheerfully in the clear waters below the bridge, building dams in the gravelly stream. It's an idyllic scene in an idyllic place. It's almost impossible to believe that only 100 years ago this beautiful place was as devastated as Bento Rodrigues and Brumadinho.

For nearly a century the valley of Forest Creek was a barren, sandy wasteland. Miners dug it over again and again after the discovery of gold in the creek bed in 1851. Both valley and creek were filled with mud and mining waste, and then torn up again by dredging early in the twentieth century. Almost every tree for miles in all directions was cut down to slab the mine shafts or feed the boilers.

Forest Creek may look like a lovely place today, but it is no gentle natural landscape. It is a complex place where the results of past human interventions and dynamic natural processes have become closely entangled. The river has gradually reshaped its bed according to the hydrological processes that govern flowing water everywhere, but the materials it works with, the sands and gravels and silts of the channel and surrounding floodplain, were deposited there by people a century and more ago. The loveliness of the valley today can be attributed to the resilience of environmental systems that endure despite repeated human onslaughts. But in the deceptively smooth contours of the land, in the mix of introduced and native plants, and even in the location of the river itself and the shape of its channel, the practised viewer can see the subtle marks of what has happened in the past.

The Forest Creek valley is one of many such entangled places. In the nineteenth century Victoria was the centre of a global resources boom, with the surging economy and environmental consequences to match. Gold from the mines built Marvellous Melbourne, while mud oozed from the goldfields to choke numerous rivers across Victoria. In those days tailings dams were unheard of. When BHP was formed as Broken Hill Proprietary in 1885 it was common practice to just let the waste flow where it flowed, and no one thought differently. Tailings covered vast areas of land in mining regions all over the world, including Australia.

Today, environmental protection laws in developed countries require mines to store their waste water on site. Modern tailings dams like the one that failed in Brazil are intended to keep mining waste out of rivers. Mining companies build ponds next to their processing plants and fill them with all the water that has been used to process the ores. These dams contain millions of cubic metres of highly toxic liquid and slurry. This form of storage is considered modern industrial and environmental best practice, and dams hold thousands of megalitres of polluted water, keeping them out of waterways. Most of the time the dams work well, but sometimes, as at the Germano iron ore mine, they give way and the accumulated toxic waste of many years is released in one catastrophic event.

Tailings dams and other environmental protection measures are a recent set of developments. In Australia they have their origins in the long history of the gold rush. Modern laws regulating the resources industry are the result of generations of struggle against the mining waste that once filled river valleys. People had a name for these waves of sand, clay and gravel that choked the rivers and blanketed the fields. They called it 'sludge', and they wanted to be rid of it. Victorian communities were some of the first in the world to successfully challenge industrial pollution. Understanding how they succeeded and why it took so long provides us with vital insight into contemporary struggles to balance mining interests with environmental values.

We first encountered sludge as part of our research into water supply for the gold rush. We are industrial archaeologists; we study the physical remains of how people worked in the past – the machines and processes they used, the raw materials they worked with, the houses and communities they built and lived in. Our evidence includes everything from the rusty bits of metal and crumbling ruins found in the bush all over Australia to the shapes and directions of rivers and

4

creeks. For a long time we've been studying the abandoned sites and ruins where people once mined for gold. A few years ago, we started to wonder about the enormous eroded pits left behind on some of the goldfields, about how they were made and who created them.

One of those pits sits at the top of Tavistock Hill in the Creswick State Forest. It looks like an abandoned gravel quarry. The hole covers more than five hectares and is up to ten metres deep in places, with steep walls cut into the surrounding clay soils and mounds of gravel covering the base. It is one of hundreds of such pits in Victoria, and not even an especially big one. It's very clear that the miners removed all the soil to get the gold, but where, we wondered, had they put the dirt? One clue lies in a series of deep crevasses on the downhill side of the pit. These gullies were drains for water to carry sand, gravel and silt into the valley below. So, the answer to our question was that the miners simply flushed all the soil into the nearest stream. We wanted to know more about this. Where did the soil go? What happened downstream? Did anyone complain? Thus began our remarkable journey into the wet, muddy, sludgy world of gold-rush Victoria.

Historian Tom Griffiths has written that part of the purpose of history is to astonish us with the strangeness of the past.[2] As we retraced the viscous narrative of Victoria's mining waste we found a story that constantly astonished us with the scale of what the miners had done – the billions of litres of water they used, the thousands of kilometres of channels they dug to carry that water, the fortunes they made selling it, and the huge quantities of sludge that moved down the river valleys. Part of our astonishment was that so much of this story seemed to be forgotten. In the popular imagination the gold rush was a brief, romantic episode of carnival that interrupted the serious business of establishing farms and building cities. There are many excellent histories of the gold rush, but there is still no definitive history of the mining industry in Victoria. The decades-long period that followed the rush, when industrial mining was the colony's economic powerhouse and much of the pollution was created, has mostly been overlooked. The environmental impact of mining is a huge void in these histories. Scholarly research on environmental history in Australia began with the timber industry and has expanded to include pastoralism, agriculture, fire, drought and flood, but it has yet to seriously tackle mining. Australians have not yet told each other the long history of mining's impact on the environment and its complex and intimate relationship with water.

The story of mining, water and the environment in Victoria is both remarkable and sobering. The state's rivers flowed with sludge for more than fifty years, closer to a hundred in some cases. In the archives we found report after report, enquiry after enquiry, petition after petition. In the digital repository for Australia's newspapers maintained by the National Library of Australia, newspaper searches revealed literally thousands of articles about sludge. This was not some short-lived crisis that passed quickly once the madness of gold fever had eased. Sludge was a chronic environmental disaster that became the normal situation for several generations.

Sludge was not a little, local problem either. Mining and mining waste affected three-quarters of Victoria's rivers. An important source of information on this environmental catastrophe is the rich archive of statistics published regularly by the old Mines Department, much of which has now been incorporated into the VicMine database. Our colleague Jodi Turnbull loaded the data on mine locations into a geographic information system (GIS) database that converted it into a map. Then she combined the mine locations with more GIS layers capturing Victoria's river systems and catchments, allowing us to compare the distribution of mines with the shape of watersheds. It produced a startling picture. From the Wimmera in the west of the state to the Snowy in the east, nearly all the major rivers in Victoria have their headwaters in the hills where gold mining took place. Mines tipped their waste into all of them, with sludge flowing south into Port Phillip Bay and Bass Strait, and north into the Murray.

The story we recount here is not just a historical curiosity. It remains important for Victorians now because mining waste reshaped the rivers and floodplains that we continue to live on, farm and manage. Understanding their nineteenth-century history helps us to make better decisions today about how we use and care for these places. The story of sludge is significant well beyond Victoria. This is a forgotten chapter in the long history of our modern attitudes to water, and in the modern identification of water itself as a commodity that could be bought and sold. These are the first stirrings of our awareness that industrial by-products constituted a form of pollution that should be controlled rather than just tolerated. Here is an accounting that weighed up the costs and benefits of different kinds of land use, and concluded that sometimes the costs of resource extraction might indeed be too high.

Nineteenth-century Victoria was a jewel in the crown of the British Empire because of the golden wealth that poured from its soil. Two per cent of all the gold ever mined in the world has come from Victoria – at least 2500 tonnes (seventy-eight million troy ounces) and probably millions of additional ounces that were never recorded.[3] Most of this gold was recovered before the First World War and almost half of it came from shallow surface deposits. Victoria is one of the world's most important surface goldfields, along with Siberia, California, the Klondike region of the Yukon, and Otago in New Zealand.

This gold-derived prosperity meant that Victorians, perhaps unwittingly, also led the way in wrestling with a new set of problems. They argued over how individuals could profit from the sale of

7

a natural benefit like water, and, if that were possible, how this market ought to be controlled. They worried about how to make a living and grow the economy while also protecting environmental amenities. They grappled with how to balance the demands of mining with the needs of the broader community. In a world where access to water is increasingly contested – and shall inevitably become more so – and where we are ever more reliant on the products of resource extraction, the environmental history of Victorian gold has never been more relevant.

Society has a short memory. There are still people alive today who remember the roar of dredges and who saw the creeks change colour when the day's sludge ran down them. Despite these echoes, the stories of the sludge that once filled Victoria's waterways have been all but forgotten. History can recover some of those stories and hold them up for re-examination. We can turn them over, peer into the corners and see that others have been here before us. In the stories of Victoria's golden past perhaps we can see hints of how we can learn to live sustainably, now and in the future, with both the water and the mining that we need so much.

1

SLUDGE

WILLIAM DUNCAN FARMED A SMALL PROPERTY near Heathcote, in central Victoria, in the 1880s. He grew wheat and oats, and grazed sheep and cattle on his 120 acres, but mine tailings were destroying his land. They had filled the creek to the brim and oozed over his paddocks to ruin the crops. When stock tried to reach the creek they got stuck in the mud, and the water was too thick for them to drink. A dozen large cattle had been lost in the neighbourhood, and others were injured when the farmers used bullocks to haul them out of the muck.

Duncan wasn't alone in his struggles. Farmers and graziers all over Victoria were seething about the sludge problem. After thirty-odd years of watching mine waste fill their waterholes and inundate the land that was their livelihoods, they had just about had enough.

At the Pink Cliffs Geological Reserve just outside Heathcote, the source of the sludge that covered Duncan's land is brutally obvious even now, a century and a half later. The blasted lunar landscape of pink, yellow and orange rock is all that remains from the decades of hydraulic sluicing that occurred here (see picture section). Much of the sluicing was carried out under the supervision of Thomas Hedley, a Canadian-born miner who managed the McIvor Hydraulic Sluicing Company. In the 1870s, the company constructed a channel, or water race, almost fifty kilometres long to bring water from the hills near Tooborac. At full capacity they used more than twenty million litres of water every day, sending massive volumes of sludge into McIvor Creek. Hedley agreed with the report from a government enquiry in 1886 that the waste water was as thick as porridge, but claimed that the shallow fall of the land meant there was no way to hold back the sludge in retaining dams and, anyway, the cost would be crippling – it was all he could do to make the claim pay as it was. He had heard complaints from farmers downstream about sludge, but he was a goldminer and, frankly, sludge was their problem.[1]

Sludge exacerbated the impact of floods. In September 1870, days and days of incessant rain at Castlemaine poured water into creeks already choked with mining debris. The creek waters rose and covered the flats of the Chinese camp, turning it into a lake and ruining stores and buildings. Bridge footings became sludge traps, forcing the floodwaters up the banks and roads. At the foot of the hill near the gaol, water rose four feet inside people's houses. Sludge from more than 300 puddling mills buried fences and destroyed gardens.[2] The reactions of locals to the drama were fairly measured – it had happened before and no doubt it would all happen again.

Sludge could also be a death trap. Two-year-old William Hamilton wandered away from home near Maldon one afternoon in November 1868. His father and neighbours spent a frantic night searching for the toddler, but by the time they found him the next morning he was lying face down, dead, in a sludge hole. In 1862, William Dow was killed at Porcupine Flat near Ararat when a shaft adjacent to his own collapsed, burying him in sludge and water. A similar accident smothered and killed nineteen-year-old William Lewis at Blackman's Lead in 1858. In 1860, Chinese miner Ah Fong was working a sluicing paddock at Fryers Creek next to one that was already filled with twelve feet of sludge. The wall dividing the two paddocks gave way and the sludge rushed in, burying and killing the young man before his brother could rescue him (see picture section).[3] There were many other sad examples of death and injury from the sludge menace.

Sludge was the dark side of the gold rush. Payable quantities of gold had first been announced in Victoria in 1851, igniting a mining boom that was heard and felt around the world. Hundreds of thousands of migrants rushed to the new colony, eager to make new lives for themselves. Few ultimately made fortunes via gold, but most found the social and economic opportunity they craved. Gold rapidly transformed Victoria from a sleepy pastoral outpost into one of the richest and most exciting places in the British Empire. The upheaval of the gold rush was the catalyst for political reforms, such as universal manhood suffrage and the secret ballot. It paved the way for widespread home ownership and underpinned the birth of the union movement.[4] The gold rush took its place among the most important series of events in Australian history. But all that wealth came at a cost, and most of that cost has been forgotten.

'What is sludge?' asked the *Geelong Advertiser* in an 1859 editorial:

> It is the Colony's share of the produce of gold-digging ... When all the other glories of a gold field are departed, the sludge will remain as a monument, in perpetuo, of wasted energies and a false policy ... The sludge is rolling down, like a lava-tide, upon the cities of Ballarat and Sandhurst, threatening to submerge them, and thus preserve them, like Pompeii or Herculaneum, for the delectation of antiquaries a thousand years hence. No less overpoweringly does the avalanche of imported sludge shoot into the seaboard towns, bearing down [on] all native industry.[5]

In 1859 Geelong was a full two-day coach journey from the source of sludge on the goldfields, and yet the threat was making its presence felt even at that distance. On the goldfields themselves the situation was much, much worse. Poor Jacques Bladier, a prominent viticulturalist, was up to his neck in it.[6] He grew prize-winning fruit in his orchards and vineyards along Bendigo Creek at Epsom. Year after year Bladier's grapes had taken top prize at the annual Bendigo agricultural show, and with them he made wine that was the toast of the colony. Now Monsieur Bladier watched helplessly as the rising sludge tide from Bendigo's mines gradually suffocated his vines and choked his fruit trees. He built levee banks five feet high to protect his orchards, but still the sludge came, flowing over the levees and making a desolate wasteland of his beautiful, verdant gardens. Monsieur Bladier's land was already Pompeii.

Sheepwash Creek near Bendigo, 1863

The landscapes of the goldfields had changed very quickly. Just a few years earlier the Bendigo valley was carpeted with green grass and dotted with beautiful gum trees, while its wattle-shaded creek banks sloped down to waterholes full of sweet, clear water. The site that was to become Beechworth had originally been a wooded plateau encircled by tree-clad hills, dipping gently towards a quiet-flowing creek in a gully. When the first miners arrived at Creswick Creek in 1852 they described the exceptional beauty of the broad flats covered with long green grass, reeds and trees, and the clear water of the creek bubbling past. But sludge brought an end to all that.[7]

'Sludge' was the colloquial word for the thick, semi-liquid slurry of sand, clay, gravel and water that flowed out of mining operations in Victoria. Farmers likened it to porridge. In California they called it 'slickens'. Experts had more technical definitions. According to the 1906 *Annual Report on Dredge Mining and Hydraulic Sluicing*, sludge was water that contained 'more than 300 grains per gallon of insoluble material', such as clay, mineral and metal. The Dredging and Sluicing Inquiry Board of 1914 decided it also included 'quartz boulders and gravels intermixed with clay and earth'. Sometimes miners distinguished between the fine liquid clay portion (slimes or slums) and tailings, which could be either the sands that resulted from crushing ore in stamp batteries, or the cobbles and boulders that were stacked in piles on alluvial claims (mullock). It all created a mess.

At times people tried to put sludge to good use. In 1859 a potter at Ballarat used sludge to make well-fired chimney pots. The specimens were described as 'beautifully smooth, of fine texture and [ringing] as soundly as a bell'.[8] Thomas Ramsay was a farmer at Newbridge who believed a little sludge on his land was a good thing. Four inches or so was ideal, so he could raise part of the original soil with the plough and mix it with the sludge. In the late 1850s, the Bendigo Sludge Commission proposed using sludge to build sixteen miles of railway embankment on the line going north to Echuca. The Eaglehawk Public Gardens (now Canterbury Park) were formed from the 'judicious use of sludge and tailings'.[9] Council workers dug out hundreds of loads of gravelly sludge from Jim Crow Creek and spread it on the roads around Daylesford. Contractors building the Geelong water supply system got gravel for their concrete from mine tailings in the Moorabool River below Dolly's Creek and Morrisons.[10] Sludge was also valued by some for the leftover gold it contained. No matter much they tried to use it, however, more and more was produced.

For miners who wanted to reprocess sludge and tailings to extract residual gold, the question of who owned the sludge was quite important. If it was stacked on a claim or held in a retaining pond it remained the property of the miner, but if it flowed down a creek it became part of the waterway and thus owned by the Crown. When it spread out over a floodplain and onto a paddock it was the farmer's problem. The 1897 *Mines Act* legislated that tailings and sludge became the absolute property of the Crown when a mining lease expired or was abandoned. The minister could then issue a licence for someone else to treat the material.

In the mid-1850s, Bendigo miners collectively discharged as much as sixty-seven million litres of sludge each day into Bendigo Creek. That meant nearly 44,000 tonnes of dirt were sent downstream from the Bendigo diggings alone *every single day*. That was the considered estimate of James Sandison, a thoughtful and observant man who studied Bendigo's sludge in detail. Sandison was a miner himself and worked the puddlers at Bendigo's White Hills. A puddler was a large ring-shaped trough in the ground, used to separate grains of alluvial gold from heavy clay soils (see figure on page 15). Ideally it was built on a bit of a slope, so water could flow in and sludge could flow out. Soil and clay were dumped in the trough, mixed with water and stirred so that the heavier gold sank to the bottom. After the water was released the gold could be scraped up and saved. The water that drained away was thick and viscous with clay. It had been turned into sludge.

Sandison's calculation of the quantity of sludge was based on his own working day, and those of his neighbours. He started with the

Horse puddling machine, Forest Creek, 1855

size of the cartloads of washdirt that miners brought to the puddler: 'a load of stuff is a cubic yard, that is a ton … those I know and nearly all do that'. He multiplied the size of the load by the number of loads a puddler could process in a day – twenty-two loads, 'a very low average'. (A contemporary of Sandison's, William Kelly, believed the average number was more like sixty loads per day.) Sandison then multiplied the daily loads by the number of puddlers operating at the time. 'There are at present', he recorded, '2000 puddling mills at work. There is a larger number of mills than that in existence, but they are not at work.' In this way, Sandison worked out the total amount of washdirt processed by the entire Bendigo goldfield, arriving at the considerable sum of 132,000 cubic yards (100,000 cubic metres) of sludge every twenty-four hours, two-thirds of it water and one-third silt and clay. He described a 'high tide' of sludge every evening around eight p.m., as the results of the day's work flowed downstream to the flats at Epsom. He knew that it travelled at a rate of 3.5 miles an hour, because he had walked alongside it.[11]

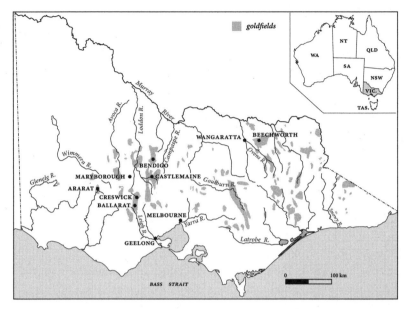

Map of Victoria's goldfields

James Sandison studied the sludge because he could see the damage it was causing, and he wanted the authorities to build a drain to divert it away from the diggings. When the Victorian government called the royal commission into Bendigo's sludge in 1859 Sandison told them what he knew about the problem. The commissioners heard a great deal of evidence, and they saw the situation for themselves. They could see that some of the sludge was trapped in abandoned diggings even before it reached Bendigo Creek:

> large areas may be seen of partially worked ground which have
> been overflowed by the sludge and rendered almost worthless ...
> [the sludge] has spread in vast mud estuaries ... in some instances
> it has risen to so great a height that the machines themselves
> have been totally submerged ... [The] sludge has attained a
> height of twelve feet above the level of the old workings.[12]

Following the commission's report a short drain was built and sludge continued to flow into the creek.

When the commissioners returned in 1861, those who lived downstream told graphic tales of the damage (see figure below). Sandison, now working as an inspector of reservoirs, was back, but he was representing his neighbours rather than his mining colleagues. We can only imagine his frustration as he described the businesses and properties destroyed in the years since he first spoke to the authorities. He described land 'covered by sludge to the extent of four feet and a half ... [It was] the finest garden in Victoria [and is now] a perfect sea of sludge, and a great many of the trees are dead.'[13] The garden he described belonged to his neighbour, the aforementioned Jacques Bladier, he of the vanquished grapes. When the sludge came, Bladier had lost dams, wells, 300 vines, ninety-four fruit trees and a business employing two other gardeners.

Sludge in Spring Creek, Daylesford, 1858

Altogether it was thousands of pounds worth of damage.[14] In the same neighbourhood, the dismayed and thirsty patrons of the Epsom hotel watched as the sludge rose as high as the windowsills.

The sludge oozed north, past the once verdant market gardens in Epsom and Huntly. Pastoralists further downstream near what later became Elmore and Rochester also claimed damages. The commissioners travelled sixty kilometres north to visit these localities and saw the situation for themselves. They recorded that

> a vast extent of pasture land is totally ruined by the sludge and rendered useless ... an area of many square miles is partially covered by the sludge to a depth of several inches. In fact the entire course of the sludge through these plains is marked by filled up watercourses and flooded pasture lands.[15]

In the 1990s, Lynette Peterson, then a graduate student in geography at Monash University, mapped how far north Bendigo's sludge had spread. She examined historical documents and scientific reports, and drove the back roads with her husband, Jim, a soil scientist. Lynette drew maps of all the places where they found evidence of sludge, which she described in precise scientific language as a 'hard-setting clay layer' containing 'angular quartz chips, pisolites, and traces of gold', found over an area of about 700 square kilometres. She concluded that 'the layer is not part of the pedogenic process of the area concerned but is in fact an allogenic depositional layer which has unequivocally altered the landscape.'[16]

The farmers Lynette spoke to had another term for those angular quartz chips and pisolites. They called it 'swamp cement'. They were very familiar with the hard, impenetrable layer of soil in their paddocks that re-formed during the winter rains each year, no matter how often they ploughed it. What they didn't know was that it was not part of the natural sequence of soils on their farms – 'pedogenic', meaning soil forming in the language of soil science. The swamp cement was 'allogenic', mineral or sediment transported to its present position from somewhere else. It was the sludge washed in 150 years ago from goldmines sixty kilometres to the south at Bendigo.

Farmers downstream from Bendigo were not alone in their suffering during the sludge epoch. In Ballarat the surrounding hills directed sludge into the valley of Yarrowee River, concentrating it particularly in Ballarat East (see figure on page 20). There it caused constant trouble for the shopkeepers along Main Road in the commercial heart of the diggings. The worst years were the late 1850s, when puddling and then sluicing directed large volumes of sludge onto the flat. The road level was raised and then raised again to lift it above the mud, and the shops on their wooden stilts got higher and higher. Eventually the road was nearly four metres higher than original ground level. By 1859, the Ballarat Sludge Commission had been charged with constructing channels to contain the waste, but still not enough was being done. Meetings about sludge attracted up to 500 people at a time and locals protested that 'unless something was done at once, the Mainroad would become flooded with sludge and water'.[17]

The sludge was eventually diverted from the Main Road shops in Ballarat and flowed down the Yarrowee River and into the Leigh River, a waterway described in the early 1850s, before the sludge arrived, as 'like a silver thread winding prettily through green, undulating country'.[18] It didn't stay that way for long. Within but a few years, sludge completely covered the river flats forty kilometres to the south at the small rural localities of Inverleigh and Shelford. In 1869 sludge transported down the Leigh from Ballarat had 'covered the low-lying lands of Captain Berthon's property a foot deep'.[19] So many complaints came from the Shelford district that it held its own inquiry in 1872. The government sent the district surveyor,

Gold mining at Black Hill, Ballarat, 1861

Henry Morres, to inspect the carnage. He reported that 'much damage has been done ... at various places along the whole course of the stream'. He found several properties 'rendered valueless' and places covered by up to two feet of mining waste. Morres talked to Ballarat mine managers about the source of the sludge. One told him that, from a mine that processed 1600 tonnes of ore each day, 'about one-half of the alluvial soil raised and washed ... is passed away in sludge'. Ballarat had hundreds of mines that processed many thousands of tonnes of ore. John Wall, an early mining surveyor, reported to the Board Appointed to Inquire into the Sludge Question in 1887 that sixty million cubic metres of sludge had flowed into the Leigh River from the Ballarat mines.[20] Twenty years later, when the Sludge Abatement Board visited nearby Shelford, the problem was just as great. The Board published graphic photos showing a thick layer of sludge coating the river flats, with fence posts and dead trees sticking out.[21]

Sludge in the Leigh River troubled Geelong residents too because it eventually ended up behind the barrier at Breakwater. Breakwater was a weir on the Barwon River just south of Geelong. It was built in the 1840s to give residents a fresh water source by blocking salt water that moved upstream. Breakwater held the sludge from the Leigh and trapped sludge flowing down the Moorabool River from mines at Dolly's Creek and Steiglitz. The Moorabool and the Leigh are both tributaries of the Barwon River, which is how the sludge ended up at Breakwater. As late as the 1930s, people living and working downstream were concerned about the impact of mining on the Moorabool. Proposals for a dredging operation at Morrisons led to protests from the owners of woollen mills along the Barwon, market gardeners and orchardists at Batesford, and the Geelong Anglers' Club.[22]

Traces of sludge at Geelong were identified in 2010 by a group of scientists from Federation University in Ballarat. They took samples from Reedy Lake, the freshwater lake above Breakwater, and from Lake Connewarre, the tidal wetland closer to the coast. They drilled into sediments at the bottom of both lakes and measured concentrations of various minerals. Arsenic was found in both places but mercury from mining was only detected above the barrier in Reedy Lake. The researchers concluded that these significant wetlands, internationally recognised under the Ramsar Convention for their role in supporting birdlife, were safe for the moment, so long as the sediments remain undisturbed; they will require careful management in the future to keep the sediments from mobilising, acidifying and releasing their toxic burden.[23] Both arsenic and mercury were used in the wool-scouring industry as well as being associated with gold – mercury for processing, arsenic as a by-product – which partly explains their presence in the lakes at Geelong.[24] In general, while we know that Victorian miners used less mercury than their Californian counterparts, not enough is known about contaminants in sludge.

Victorians far and wide protested the effects of sludge. At Creswick in the 1850s sludge accumulated in the streets and made them impassable. Shopkeepers complained that it was driving customers away and damaging business.[25] Heathcote residents found that 'a great accumulation of sludge had proved a sad interference to mining operations and a serious damage to private property'. Four hundred of them put their names to a petition on the problem in 1860, but sludge continued to flow into McIvor Creek and the Campaspe River.[26] Mines around Maryborough filled local creeks with sludge and sent it down Bet Bet Creek, while mines at Clunes sent theirs down Tullaroop (then 'Deep') Creek as far as its junction with the Loddon River at Eddington, leaving eight to ten metres of sludge in what had been deep waterholes. Mr Outtrim, the shire secretary and, incidentally, manager of one of the local mines, told the 1886 Sludge Inquiry that he'd never had any official complaints, although he grudgingly admitted that 'farmers are continually complaining that each flood that comes down washes the silt further over their flats'.[27]

At Carisbrook, about fifteen kilometres upstream of Eddington on Tullaroop Creek, farmers were so outraged that they sent a petition to the minister for lands in 1887. They had the same problem as farmers at Shelford below Ballarat. Sludge from quartz crushers upstream at Clunes was flowing down and inundating their land. Their farms were covered by up to half a metre of sands that were 'highly inimical to vegetable life', and the water was too contaminated to drink or even wash with. Many had been affected, some losing as much as forty hectares of agricultural land to the sludge plague.[28] Members of the Sludge Board visited the area in 1886 and agreed:

Deep [Tullaroop] Creek … is much silted up, and filled with quartz, sand, and slum, the old creek bed … being filled up

level, and the water passing along a new channel ... Along the
line of the [Loddon], the trees in the watercourse have died,
being killed, it is stated, by the water discharged from the
Clunes quartz mines, and the injury to the low-lying lands is
shown by the evidence to be very serious.[29]

People far downstream on the Loddon around Kerang worried that
the mineralised water pumped from Maryborough's deep lead mines
would contaminate the drinking water they drew from the river.
Maryborough folk dismissed these 'northerly situated grumblers',
for 'it is well known that many of the miners drink the water in the
claim'.[30] Miners at Creswick drank water from the deep lead mines
too, according to William Guthrie Spence, secretary of the Miners'
Association. Spence had started his illustrious career as a pioneer
of Australia's labour movement in the mines beneath Creswick, so
he knew them well.[31] People downstream from the miners Spence
described, however, weren't so keen on the sludge getting into their
water supply. Clunes residents took their water from a weir on Birch
Creek below the Creswick deep lead mines. In 1881, landowners along
Birch Creek went to court to stop the mines pumping sludge into the
creek. Responsibility and blame, however, were passed around in a
circle. One of the main users of water in the Clunes weir was the Port
Phillip and Colonial Gold Mining Company, which dumped tailings
lower down the catchment into Tullaroop Creek. Maryborough miners told the 1887 Sludge Board that it was really the Clunes miners
who were filling the waterholes in the Loddon River at Eddington,
and it was the Clunes mines that the Carisbrook farmers blamed for
their woes.[32] They all blamed each other and no one was happy.

Sludge in the water supply was a concern for many downstream
towns. In Gippsland, sluicing on the Mitchell River and its tributary
the Dargo produced water that was 'greatly discoloured ... clayey in

appearance and quite unfit for drinking or domestic purposes'.[33] At times the sludge reached as far as Bairnsdale and Lake King, causing great consternation at the Bairnsdale Water Trust.[34] Mining on the Thomson River affected the quality of drinking water at Sale when sludge from the mines at Walhalla flowed downstream. When there wasn't enough rain for the sludge to flow, tailings from the Long Tunnel Extended Gold Mine blocked up the creek below the stamp batteries that crushed the ore, stopping work until the creek could be cleared.[35] The Goulburn Valley towns of Seymour and Nagambie worried that dredging upstream would affect their water supply – and they were right. By 1910 sludge was discolouring the water at Seymour and there were fears that the Goulburn Weir would silt up. The weir provided irrigation water for the Waranga Basin and siltation would significantly reduce its capacity.[36]

The concern that sludge could fill dams and weirs far downstream was not alarmist. Laanecoorie Weir, for example, accumulated three metres of silt in its first twenty years of operation. It was built on the plains of the central Loddon Valley in 1891, and by 1908 locals were telling the Sludge Abatement Board about the muddy waste that was filling it. Laanecoorie is just below Eddington, so the weir collected sludge from creeks draining Maryborough, Clunes, Creswick, Daylesford and Castlemaine. When in 1909 a great flood breached the wall of the weir and sent a violent pulse of water and sludge down to Newbridge, the receding waters left a wasteland of slush and slime.[37] Today, scientists and water managers estimate that Laanecoorie lost more than half its capacity in its first four decades of operation.[38] Even though most of the mining in its catchment had finished before the weir was built, the sludge was still working its way downstream to become trapped behind the embankment.

The sediment that filled Laanecoorie still travels down the rivers as 'sand slugs', bars of sand and gravel that fill the riverbed.[39] Every time there is major rainfall the floodwaters push the sandbars further downstream. Sand slugs change the ecology of rivers in several ways. When they're stable, sandbars fill cracks and hollows in the riverbed to create a shallower, sandy bottom. As a result, the river has fewer deep pools and a reduced habitat for aquatic plants and animals. When the sand slugs are moving during floods, they scour the riverbed, scraping it clean and leaving an altered environment behind. Joseph Reed, town clerk in Creswick in the 1880s, described the sand slugs he had seen in the three decades he lived there: 'You would find, in some places, the old bed of the creek still, the silt shifted every year in every flood ... The old debris, that settled in the creek up [sic], is being shifted down every year. The accumulation of quartz tailings is shifted lower down the creek with every flood.'[40] Even today, sand slugs from the Beechworth and Bright districts are still moving down the Ovens River and its tributaries, slugs from Castlemaine and Daylesford sit in the Loddon River at Newstead, and sand from the Ballarat mines is moving down the Leigh River towards the Barwon.

Sludge was not only troublesome and undrinkable, it could be positively dangerous to humans and animals. Horses and cattle died from an accumulation of sand and silt in their bellies after drinking sludge-filled water. Loddon grazier Tom Beveridge reported that 'a number of my neighbours' stock have died, and I believe it is through nothing but drinking this water ... It is an accumulation of sand ... in the intestines.' Also on the Loddon, Alexander Clarke described mud 'of such an adhesive character' that it stuck to his cattle 'like cement'. He had seen 'cases of the hair coming away with it when it at last peeled off'. Cows wandering down to drink at Campbells Creek in Castlemaine routinely got bogged and half-drowned.[41] Sludge-filled creeks around Heathcote meant that cattle coming down for a drink got stuck in the mud and were injured, sometimes fatally, when they had to be pulled out.[42] In 1865 four bullocks were found buried up to their necks in a large sludge-flooded paddock by the Epsom road near Bendigo. It proved impossible to get the animals out alive, so they were killed, and horses hauled their carcasses to the slaughter yard.

People, as we've mentioned, also drowned in sludge. Terence O'Brien had been drinking at the pub late one night in 1869 when he fell into a sludge-filled pit in High Street, Bendigo, and was found dead the next morning. A coroner's jury of twelve 'good and lawful' men blamed the local council for leaving a sludge pit unprotected in the middle of a public road.[43] Three-year-old James Rodda drowned in an Eaglehawk sludge channel in 1879, while the elderly James Williams was found drowned in a sludge dam at the Lord Nelson Mine at St Arnaud in 1910.[44] In 1901 Julia O'Reilly was suffering insomnia when she stumbled into a sludge-filled creek near Invermay, near Nerrina (Little Bendigo), just north of Ballarat, a region well known for its sludge problems. The creek had 'a couple of feet of sludge but very little water', despite the fact that it was early October, when it should have been full of spring rain.[45] O'Reilly was rescued by the local policeman who had been searching for her. Invermay was one of the places visited by the official Board Appointed to Inquire into the Sludge Question in 1886. They found that companies operating in the Dowling Forest were sending tailings from their stamp batteries directly into Burrumbeet Creek.[46] The quantities of sludge from working and proposed companies was so great the board members were concerned that Burrumbeet Creek would be 'obliterated' and Lake Burrumbeet would suffer permanent damage.

Near Tarrawingee, in 1876, a toddler was lost in the sludge for two days. Willie Whittle, just two years old, strayed into the sludge near his home one winter afternoon and was only found because his tracks were visible in the yellow mud. The next day searchers following the trail 'were rewarded by seeing the little fellow lying down at the foot of a tree, having fairly laid down to die – being in a bed of sludge with three inches of water round him'.[47] Tarrawingee was even more notorious as a sludge hotspot. It was and still is a region of rich agricultural lands, on the flats of the Ovens River, east of Wangaratta. For sixty years mountain creeks drained the Beechworth mines and dumped sludge on farms downstream. It lay thick on the ground for miles in every direction, a slick pool of mud to entice a small boy.

The year before Willie Whittle strayed, local councillors had estimated that over 10,000 acres of land had been destroyed in the district. The *Ovens and Murray Advertiser*'s account of the situation described a catastrophe: 'For miles along the creek the sludge and mullock is being spread over the farms, ruining land, destroying crops, annihilating fences, and bringing about dire destruction. In some places, land once grain yielding and valuable is covered from a depth of from six to ten feet.'[48]

The newspaper feared the fertile flats would become a 'barren, howling wilderness'. Because the natural river channels were choked with mining waste there was nowhere for stormwater to flow. After heavy rains 'the water spread all over the plains, invading and deluging farms and gardens, destroying property, and driving the unfortunate inhabitants out of their houses and homes'.[49] Sludge could also accumulate to great depth. When the Pocahontas Gold Mining Company was formed at Amherst in the early 1900s, for example, workers found a complete tree buried in the sludge at a depth of fifty feet. On the Loddon River, waterholes filled with twenty-five to thirty feet of sludge. At Eldorado, the Cocks Pioneer Gold and Tin Sluicing Company used coarse tailings and green branches to form a retaining dam about ninety feet high.[50]

When the Sludge Board visited Wangaratta in December 1886 they heard from twenty-five local people, most of whom were farmers. Witness after witness described the place prior to the discovery of gold as 'a sort of chain of waterholes in summer time, and when the floods came down it became a river', with excellent fishing, rich market gardens and orchards, and lush grass on the river flats. And now? A layer of sludge half a metre thick lay across the river flats in some places, while in others fences over a metre tall were buried in sludge. One farmer's land was 'as barren as a sea beach'. A bridge across the creek that someone on horseback had once been able to ride beneath was now so silted up that 'you can hardly crawl through'. Fishing had been destroyed because 'the river is thick like gruel', and the grass that struggled to grow on the inundated land 'is no good, it is sour, the cattle do not seem to care for it at all'.[51]

The sludge flowed down Hodgson Creek from the diggings south of Beechworth to the junction with the Ovens River opposite Oxley. Farmers there testified that 'the Ovens is silting up … the deep holes in the river are as level as this table now for from half a mile to two miles … from the mouth of the sludge channel'; the original bed of the river had risen three to five feet.[52] Far below the junction, landowners along the Ovens reported thick deposits of slimy yellow clay on their properties. Twenty kilometres downstream from Wangaratta; at Boorhaman, farmer James Pratt was adamant that 'at times the river is quite thick, and unfit for any purpose'.[53] He was sure it was from mining and no other causes because on Mondays and Tuesdays, after the miners' Sunday day of rest, the water would run clear. The silt reached as far the Murray River, according to reports Henry Parfitt had heard from people at Yarrawonga. He had seen with his own eyes six to eight inches of sludge on the river flat.[54]

In general, Beechworth witnesses were reticent about the damage. Most of them were miners, including John Fletcher, the shire secretary, who went so far as to say he had heard no complaints. That seems highly unlikely, given that his shire had been in dispute with the North Ovens Shire for at least fifteen years about sludge. When pressed, Beechworth mining surveyor Henry Davidson reluctantly agreed that 'I have seen, in very dry seasons, about the end of March, the bed of the Ovens River, below Wangaratta, covered with a hard cake coating of sludge, and after floods I have seen this slum lying on the flats around the lagoons'.[55] Davidson was quick to insist that 'There has been serious damage in the past, but I am not aware of any present damage at all'. We can imagine the straight faces kept by the members of the Sludge Board as they listened to such denials. They had not yet heard the testimonies of the Tarrawingee farmers, but the train they had travelled on to Beechworth passed right through the worst of the affected region.

One of those who testified about – and protested relentlessly against – sludge at Wangaratta was Hopton Nolan. He had been a miner in the first rush to Beechworth in 1852, and later he took up land on the flats at Tarrawingee. Nolan prospered as a farmer, grazier, orchardist and viticulturalist. He ran the town's centrepiece, the Plough Inn, kept a store, and at the time the Sludge Board visited he was president of North Ovens Shire. Nolan owned land on both sides of Hodgson Creek and knew all about the damage suffered in the district; most of his own farm was directly affected. He told the Commissioners that to prevent damage he had built a bank a metre high around 'the bulk' of his property: 'If I had not it would have been of very little use to me.' By then, sludge had already destroyed a belt of large and ancient trees along the creek: 'I have known it to destroy gigantic trees from one foot to two feet and three feet through where it has sludged up … not only one tree, but belts of it.'[56] Nolan and his neighbours at Tarrawingee, including Henry Parfitt, later played a prominent role in driving changes in government policy – we'll return to their story in Chapter 6.

Over the ranges at Yackandandah and beyond, miners were filling rivers with sludge. Two companies with claims on the Mitta Mitta River, the Pioneer Mine and the Union Gold Sluicing Company, used hydraulic nozzles to blast away alluvium up to thirty metres high. As the Sludge Board ultimately learned, there were six million cubic metres of deposit still available for the two companies to process, enough to fill a river sixty metres wide, one metre deep and 100 kilometres long. Their response to that – that 'the future evil ... [was] one for serious consideration' – seems rather mild.[57] The digging and sluicing operations at Sandy Creek, a tributary of the Mitta Mitta were 'so extensive' that the valley was 'elevated several feet – to such an extent that the present course of Sandy Creek is an artificial channel cut through from three to four feet of earth deposits, having their origin in gold washing'.[58] It was the same at Staghorn Creek, where the surface at the turn of the century was 'several feet above the original, the difference representing the deposits of many years from the varied alluvial workings'.[59] Both valleys drew the interest of bucket dredgers who, in the early years of the twentieth century, reworked the river flats to extract the gold that remained in the sludge left by their predecessors.

The government and the miners were hostile to the suggestion that they were doing something wrong. When Jacques Bladier and his neighbours at Epsom and Huntly presented a claim for £7600 in damages to their market gardens, the government was justifiably alarmed. They realised that 'similar and equally valid claims of the same nature will soon follow from almost every part of the country ... it may well be doubted whether the whole revenue of the country will, in a few years, suffice to meet all claims of this character.'[60] After some thought on the matter, the commissioner of public works informed the claimants that the government 'does not consider the claims in question to be legitimate demands upon the

Government'. Any such damage was simply the inevitable result of proximity to the goldfields. The commissioner advised Bladier and the other fruit growers to build larger earthworks.[61]

When frustrated landowners sought redress in the courts, there was general community outrage. But surely it was the miners who were in the right? They risked their money to discover good, honest gold, producing jobs along the way. Hadn't they built Victoria through these pioneering efforts? Why stymie the industry when it was so obviously a force for good?

This view was repeatedly expressed in parliament over the years. '[Those who have] expended a large sum of money in getting their particular works into a fit state to enable them to make some money – if they are likely to be interfered with a most serious injustice will be inflicted upon them,' thundered Alfred Billson from Beechworth. Billson was echoing the sentiments of parliamentarian James Campbell, who asserted in 1885 that Creswick was 'the great alluvial gold-field of Victoria – there was no part of the country in which, within the same area, such vast interests were concentrated – and, therefore, it would be the profoundest pity if anything occurred which would prevent the amicable agreement which had been come to between the mine-owners and the landowners of the district'. Campbell, in turn, was reiterating a view proffered several years earlier by his colleague Major William Collard Smith from Ballarat, who told the House that 'To give him [the landholder] the power to inflict a great wrong upon hundreds of shareholders and hundreds of miners, without a word in explanation or defence, is a power that should not be left in the hands of anyone.'[62]

Campbell and Smith were both agitated about cases that had arisen around Creswick. Landowners with properties downstream from rich deep lead mines were seeking injunctions to stop the damage caused by the sludge flowing over their land. In 1881, Murdock

McKenzie claimed that waste from the Madam Berry mine at Creswick made the creek on his land unfit for watering stock and buried springs under a thick layer of sludge. To the annoyance of the press, who declared these restrictions to mining 'not [a] very defensible policy', McKenzie's claim was supported by the judge, and his injunction granted. The landowner got financial compensation from Madam Berry and mining continued – along with the production of sludge. In 1885, another group of Creswick landowners once again sought compensation and the matter went before the courts, the press and parliament all over again, with similarly fruitless results. Cynicism and opportunism, however, abounded on both sides. The complainant in the 1881 injunction case, Murdock McKenzie, had a mining claim of his own, while one of the offended miners in an earlier injunction at Maryborough was none other than Thomas Outtrim, shire secretary and witness at the sludge inquiry a few years later, where he tried to convince the inquiry that there had been no official complaints about sludge in his district.[63]

And so, miners continued pumping their sludge into creeks and rivers. In fact, they got better at it after bucket dredges were introduced to Victoria from New Zealand around 1899. These floating factories dug up massive deposits of low-grade ores from river valleys and spurted waste out behind them. Between 1905 and 1913, more than thirty bucket dredges in the Ovens Valley alone worked over 1600 hectares of land, producing fifty million tonnes of debris.[64] In the Loddon Valley, dredging at Guildford continued to pour effluent into the river, while dredges also started up in many other mining areas. Jet elevators and pump hydraulic sluices were further, widely used variations on the theme of using water – and lots of it – to wash away the soil and separate out the gold.

A supporter of the gold mining industry described the effect of dredging on rivers in the *Ovens and Murray Advertiser* in 1903. They

failed to see any problem with it at all. 'Dredging is principally carried on in the beds of the running streams,' they noted, and

> as a general thing no new matter is deposited in the water
> course. The gravel, sand, soil, etc. are only transferred from one
> position to another. True the water was temporarily discol-
> oured, but that is inseparable from the industry ... it is not at all
> apparent how the discolouration of water with soil of the same
> nature as that through which it flows, and which forms its bed,
> can be termed 'pollution'.[65]

As the writer indicates, damage from dredging was twofold. First,
dredges were yet another source of the sludge that was carried downstream, choking rivers, covering flats and making water undrinkable.
It might have been 'the same nature as that through which it flows',
but it was sludge nonetheless. Thomas Graham, president of the
Ovens Anti-Pollution Association, provided another interpretation
of this 'temporary discolouration'. In Reedy Creek, below the dredge
at Eldorado, it had been the case for months, he complained, that 'the
bed is covered or filled with this slimy mud and absolutely no water
is available for stock'.[66]

Second, a new and greater problem was caused by dredges working directly in river channels: 'the beds of some of the streams are
being turned upside down in the course of dredging operations.'[67]
As the early dredges could work to depths of eight metres below the
surface, the river channels and adjacent banks were being churned to
an unprecedented extent. The effect was stark. In 1914 the Dredging
and Sluicing Inquiry Board compared two sides of the Ovens River,
one bank dredged and the other as yet untouched. 'The right bank ...
has been converted from its original fruitfulness to a bare shingle
beach, which apparently it must remain in perpetuity, while the left

bank, which consists of a deep loam, [is] well-clothed with fine timber and shrubs.'[68]

Dredging defenders claimed that industrial process improved the land when it reworked areas that had been previously mined. There was some justification of this as the old diggings were 'quite valueless for grazing, agriculture or even for gold mining' because of the 'destruction of the original timber and vegetation, the honeycombing of the land with shafts and drives, and the partial upturning of the surface'.[69] Dredging levelled the ground, filled the holes and removed abandoned debris. The resulting surface could potentially support plantation timber or scrub, but it was no longer much good for cropping. In an area below Bright that had been dredged extensively, the damage was still plain even two decades later: 'Huge piles of mullock and gravel lie along the roadside and are all that remain of once rich paddocks. The river has been banked up and silted with the refuse from the dredges that tore down the stream.'[70]

By the end of the nineteenth century, sludge and mining debris could be found in three-quarters of Victoria's major river systems. It flowed south into Bass Strait and drained north into the Murray River. Although gold was running out, the sludge problem was only getting worse. As the search for profits became more and more desperate, miners turned to bigger and more efficient machines that used and polluted even more water than before. As we were finding out, water supply was the key to the whole unfolding catastrophe.

2

MINING

WATER SEEPS FROM THE GROUND IN A GULLY below the farmhouse, trickling down the steep hillside in a culvert beside a farm track. From there it flows down to the flat where the valley floor is already boggy and green with bracken, gorse and ferns. Here an underground aquifer breaks through the hillside and joins other springs emerging nearby. The waters collect in a channel and run placidly down the stream, around the hills, through a pine plantation and past more farms. A few kilometres downstream the flow becomes a substantial mountain stream, running clear over a shallow bed of pebbles and sand. Only the steep, mounded creek banks hint at the human industry involved in their shaping. This is not a natural mountain stream at all. Its banks were built up long ago from soil excavated to create the channel in which the stream now runs. Further downstream the mounds become higher and the channel deeper and wider, with grassy banks that stand out above the surrounding land and keep the water in place.

The valley is high up in the ranges near the little town of Stanley. Now a quiet, pretty place of orchards and berry farms, it was once a booming mining settlement known as Snake Gully (see picture section). When miners arrived in the early 1850s they cut down trees and dug up the soil, creating the disturbed ground in which gorse and bracken now flourish. These same miners also discovered the springs seeping from the hills, and people have been fighting over them ever since. Over the next two decades, arguments over Stanley's water played a formative role in establishing the principles of water allocation that still shape Australian water management today. In fact, over the last few years, access to Stanley's water has again been contested, in a bitter wrestle between local farmers and large multinational corporations. The water, prized for its purity, is now hauled away for bottling and sale in urban cafes. During the gold rush it was valued for its reliable abundance and diverted away in channels. The miners who built the channels called them 'water races', and they were a core, indispensable feature of mining operations.

We came across the springs at Stanley when we were looking for the source of water used for mining further down the ranges, west of Beechworth. It was late September and the almond orchards were blooming. We spent a morning exploring the deep cuts in the hills produced by sluicing more than a hundred years ago. We found one that was nearly fifteen metres deep in places, with steep vertical cliffs that plunged from the roadside into the pit below. It is a huge hole that runs for more than ten kilometres along Three Mile Creek and its tributaries. In total, the pit covers over 250 hectares of river valley. This land was once home to hundreds of people. Historic maps show the locations of miners' huts, a church, a school and Chinese market gardens.

All of that is gone now, washed away in the search for gold. Those homes and gardens were replaced by the deep pit, crowded with blackberry and gorse, that we have come to inspect. In some places the pit has a more open mix of mature stringybarks and manna gum, but for the most part it is almost impossible to push through the scrub. A simpler way to see the full extent of the site and sense the contours of the old diggings is by using aerial photography. Google Earth makes this much easier than it once was, but all the trees and shrubs in the way meant it was little help to us, so we turned to a remote-sensing technology called LiDAR, which stands for light detection and ranging, and uses laser beams from aircraft or drones to measure and record ground surfaces in precise detail (see example below). The resulting images effectively erase vegetation, creating what looks like a high-resolution, three-dimensional model of the surface terrain. LiDAR imagery is still expensive and hard to access, but in this instance we were lucky. In recent years, the Victorian government has commissioned LiDAR surveys of all the major rivers and streams in the state, and through the help of colleagues we were able to study the data. With LiDAR we could clearly see the sharp edges of the pit, the creek running along its base, and all the ditches and mounds left by the miners.

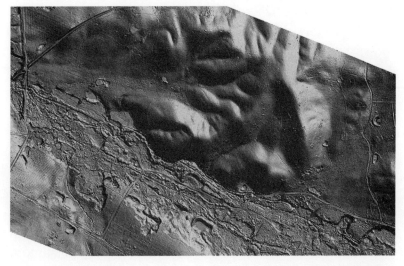

LiDAR, Three Mile Creek

Fieldwork on the ground is still important, though, and sometimes it yields surprises. Studying the LiDAR images closely, we could see faint lines like scratches running across the bottom of the pit alongside the creek. Some ran for hundreds of metres. They were too straight to be natural, so we suspected they might be the races cut by miners. Our colleague Jodi Turnbull layered the LiDAR images in GIS software with old maps of the mining claims and water licences held by the Victorian Public Records Office. Then she added aerial imagery and spatial data for the modern road network so we could physically locate these 'scratches' on the ground. At the site we used the georeferenced map loaded onto an iPad and found one of the features easily. The channel was so covered in ferns and bracken we would never have found it otherwise, but there it was: a tiny, narrow, straight ditch. It looked just like all the other water races we had been following, but because Jodi had layered all the old maps in GIS we knew that it wasn't one of them. This race didn't supply water like the others. It was, rather, one of the sluice boxes used in the mine itself to wash the gold from the soil. Originally, the sluice box would have been lined with slabs of timber, but those were long gone now and only the earthen channel remained.

This sluice box had belonged to John Martin Dietrich Pund, one of many Germans on the Ovens goldfield, and among the most successful miners in the area. He was a sail-maker from Hamburg who worked his passage as a crewman on the *Cesar Godeffroy* to Adelaide in 1854, and journeyed to Beechworth a year or two later. In the early days, Pund worked alluvial claims in the district at Yackandandah and Nine Mile Creek at Stanley. Then in the 1860s he moved over to Three Mile Creek – now known as Baarmutha – and began a successful fifty-year career as a water merchant and hydraulic sluicer, during which he recovered almost 25,000 ounces of gold. He died a prosperous man in 1915.[1] We gradually came to know this Mr Pund and his operations very well.

On the day we visited Pund's sluice, water from recent rain was cascading over the cliff edges in tiny waterfalls. The running water was much like the process goldminers would have used to work the ground in the nineteenth century. They built races to carry streams of water to the top of the working face. Then, as water poured over the edge and down the face, it loosened the soil and sent it tumbling to the bottom. Channels and wooden sluice boxes at the base of the cliff collected the water and washdirt. The water then carried away the lighter soil and the heavier gold sank to the bottom of the sluice. Miners started at the lowest point of the claim by digging channels across the ground and shovelling in the washdirt. As they gradually worked uphill, the cliff face got higher and higher and they could run the water over the edge from above. Australian miners called this manner of working 'ground sluicing' (see picture section and photos below and on page 40).

Ground sluicing

Box sluicing, c.1861

Victorians learned this method from miners with experience in California and Cornwall. The Cornish knew it as 'tin streaming', and they had been doing it since Roman times to work the rich tin deposits of the Cornwall region.[2] Similar techniques were used in the old mines of Spain and Germany. Sluicing at Beechworth was described in the 1860s by Peter Wright, a district engineer for the local water supply, who had extensive experience of mining in both California and Australia. In his account, a sluice consisted 'essentially of an inclined channel, through which a stream of water flows' in order to break up the earth being conveyed into this channel, 'carrying the lighter materials away, and leaving gold, tin ore, &c., behind'. Wright explained the difference between box sluices, 'which are raised above the bottom, into which the earth must be elevated by manual or mechanical power', and ground sluices, which were 'sunk in the bottom, into which the earth is conveyed by a stream of water'.[3]

Stanley and Beechworth were hotbeds of sluicing activity in Victoria. The first recorded sluicing operation in the colony was established just a few miles up the road from John Pund's works on Three Mile Creek. It began in the autumn of 1853 when young Irish miner John Henry Reilly constructed the colony's first water race to supply his claim.[4] Born in Dublin around 1828, Reilly trained as a civil engineer before gaining his mining experience in California in the late 1840s, where he helped build water supply systems on the Stanislaus River in Tuolumne County, working what the Californians called 'placer deposits'. Reilly was one of 100 shareholders in a cooperative company that constructed the Columbia Ditch in 1849, a channel system that eventually grew to be 186 miles in length.[5] Reilly saw that similar methods could work in Victoria, and he built a race as soon as regulations allowed it.

Others in Beechworth brought the mining skills necessary to complement Reilly's engineering expertise. Experienced Cornish tin miners were already at work in the district. Just before Reilly applied for his claim, *The Goulburn Herald* reported that the black sand found in local creek beds had been identified as cassiterite tin and a company was forming to work it on the cost-book system.[6] Although the report doesn't explicitly identify the company as Cornish, there are clues that it was. Few but the Cornish would have had the experience to identify black sand as tin, and only the Cornish financed their mining companies on the cost-book system. What's more, the company proposed to 'dig trenches and sluice the washing stuff', suggesting these entrepreneurs were experienced tin streamers. Nothing more was heard of these tinworks, so we may suppose that the members of this company were among those eager to use Reilly's water to win gold.

When word of gold discoveries in Victoria and New South Wales were first announced in 1851, eager Cornish miners working the copper mines in South Australia were among the first to arrive on the embryonic goldfields. Nearly 1000 men left the town of Burra alone in the first twelve months after gold was discovered. Many more Cornish miners came to Victoria via California, or direct from Cornwall, and Victoria very quickly became the primary destination for Cornish emigrants in the 1850s. John Phillips, for example, was a mining surveyor at Muckleford in the 1850s whose father had spent fifty years as a tin streamer in Cornwall.[7] The two men who discovered the Welcome Stranger nugget at Moliagul in 1859, John Deason and Richard Oates, were both Cornish. Deason had worked as a boy in the tin mines of Cornwall and the pair worked a puddling claim at Bendigo, so they knew a valuable chunk of rock when they saw it.[8] All of Victoria's goldfields had a significant Cornish presence, including Ballarat's own 'Cornish Town' and a 'Little Cornwall' in Bendigo. Early in the rush some 4000 Cornish miners went to Mt Alexander (now Castlemaine), and by the late 1850s seventeen per cent of the mining population in Bendigo was Cornish.[9] With that kind of expertise seeping through the colony, ground sluicing spread quickly to all the diggings.

Chinese miners also developed a reputation for being skilled ground sluicers.[10] These miners were closely associated with many alluvial goldfields, including those around Braidwood and Kiandra in south-eastern New South Wales, the Buckland River diggings in Victoria and the goldfields of northern Queensland. In later decades they were also prominent on the tin fields of north-east Tasmania. Some may have brought sluicing skills with them to Australia because Chinese miners were working alluvial tin deposits in Malaya from 1850. We can only speculate about this, because so far no one has researched links between the Chinese in Victoria and those on the Malay tin fields. But what does seem clear is that sluicing was already underway in Victoria when large numbers of Chinese miners started arriving in 1854.

Sluicing was quickly adopted on all the fields where there were substantial gold deposits in ancient river gravels, a topography with enough hills for a gravity-based system to work effectively, and, crucially, enough water to feed the sluices. Sluicing used *enormous* quantities of water. Even small claims of an acre or two could consume up to a million litres each day.[11] A claim at Fryerstown owned by James Symes and his brother Mathew was authorised to draw sixty-eight million litres of water per day from the Loddon River.[12] This optimistic licence turned out to be twice the total flow of the river but the Symes brothers were not deterred. At Creswick, the Humbug Hill Sluicing Company used 1350 litres of water per minute to work their ground. The district mining surveyor James Stevenson described the activities of the party:

> The ground was washed from the surface to the bottom – a depth of thirty feet, the lower ten of which were a soft clayey red reef, and had to be thrown up into the sluice streams. The mode of working adopted was first to cut a face on the ground, and then to turn on the water along its base. Thus the water assisted in cutting down the ground, and frequently blocks of from twenty– fifty tons were so taken down. The shifts were six hours on and twelve hours off, and the work was kept going night and day.[13]

The ground was poor, yielding less than eight grains (half a gram) of gold per cubic yard, but the large volumes of dirt sluiced offered a decent return of eleven shillings per man per day.

By the mid-1880s the sluicers who followed John Reilly's lead at Beechworth and the surrounding district were collectively licensed to use more than 450 million litres of water per day. They diverted water into thousands of sluice boxes, and dozens of waterwheels that were used to drive pumps and winding machinery.[14] John Pund was one such sluicer and he was astute in his use of water to work his claims. In 1865 he returned to old, previously worked ground along Three Mile Creek and started combining a number of small leases. By 1887 he held twenty-four hectares under mining leases and miner's rights, and used hydraulic hoses to sluice alluvial faces up to eight metres high. Long-boxed 'tail' races, some up to 1000 metres in length, ran down the valley and emptied into Hodgson Creek. Pund used over four million litres of water per day during the 1880s and increased that to 6.5 million in the 1890s. He reused the 'tail' water that flowed down from his own claims upstream, and those located below him reused the water in turn (see picture section).[15]

By the 1870s men like Pund had advanced from ground sluicing to hydraulic sluicing. Rather than just running the water over a cliff face, or shovelling the washdirt into channels and sluice boxes, they directed the water from their races into iron pipes that gradually narrowed and ended in a bronze nozzle like a fire hose (see figure on page 45). The hydraulic 'head' built up immense pressure, and water jetted out of the nozzle with enough force to dissolve entire hillsides, sending the gravel crashing down. Peter Wright described the process, with a concern for its safety implications:

> In the hydraulic system the earth is got by means of a jet of water brought to bear on the face of the cutting by means of a flexible hose with a nozzle like a fireman's branch ... When the face of a cutting exceeds twenty feet in depth, it has to be carefully watched when men are under-cutting it, and when it exceeds thirty feet in depth it is unsafe with any watching. On the other hand, I have seen cuttings nearly sixty feet in depth safely wrought with the jet, the workman directing the water being able to stand at a safe distance from the face.[16]

Hydraulic sluicing, Pioneer Claim, Mitta Mitta

Hydraulic sluicing, Omeo, c.1880

Safety, however, was not always a priority. Mining was a dreadfully dangerous industry, and men were frequently maimed and killed by collapses. In the ten years from 1874 to 1883, 632 miners were killed in Victoria and many more were badly injured.

Most deaths occurred in falls or explosions in underground mines, but earth collapses on alluvial claims were also frequently fatal. Thirty-year-old Richard Martin, sluicing the hills south of Creswick, was killed when several tons of earth fell on him in the spring of 1860. He left behind a wife and four children. Martin was one of at least seven men killed by earth falls in this locality alone during the 1860s and 1870s; the others were Chinese, and included Fun Wagh, Yung Lan, Ah Hik, Ah Luke and Fun Gwan.[17] In this period, Chinese and European alluvial miners were killed at the rate of about one man per thousand each year. The chief inspector of mines, Robert Brough Smyth, observed in 1875 that accidents were common among alluvial miners because they used little or no timber to shore up their sluicing works, and it was, he reported, impossible to control the 'transitory and unimportant operations' of these small parties whose reckless behaviour endangered their lives. As far as the Department of Mines was concerned, it was basically the miners' own fault if they were smothered and killed.[18]

Sluicing was the biggest consumer of water on the goldfields but other mining methods used plenty of water as well. At the start of the gold rush, panning and cradling were the most common techniques. There were generally three people working each cradle: one to rock it from side to side, another to shovel and beat the clay and a third to bucket water from a creek. Most of the 40,000 miners on the goldfields in 1852–53 needed water flowing in a creek, or at least a chain of water-holes, to recover gold from their small claims. At Mia Mia Creek near Amherst in 1858, 500 Chinese miners formed a bucket line to collect water for surface washing. In the absence of water, miners stockpiled and guarded their washdirt while waiting for rain. Alternatively, they could haul the dirt to a water supply. At Bendigo in the late summer of 1853, for example, many carted their earth a distance of thirteen kilometres from Golden Gully to wash it at Sheepwash Creek.[19]

Puddlers also needed plenty of water – thousands of litres per day. Some miners used a pipe with a T-handle as a plunger pump for raising water to the puddling mill; others used a small steam engine. Between them, James Sandison and the puddlers at Bendigo in the 1850s used more than sixty million litres of water each day – when it was available, that is.[20] The problem with puddling, though, was that it was not very efficient at recovering all the gold. It was well known that fine particles were carried off in the sludge and that some kind of method, such as hides or blankets, was needed to catch them. Mr Gabrielli was a London stockbroker who toured the goldfields in 1858. One day he fell into the sludge at Bendigo and was only rescued with some difficulty. His clothes were ruined but he salvaged his waistcoat, which was so stiff with sludge that it stood up by itself. Later, when washed, the garment was said to glitter with particles of gold dust.[21]

Sluicing and puddling are both methods for working alluvial deposits of sand, gravel and clay from ancient riverbeds. The deposits sluiced at Beechworth and puddled at Bendigo were fairly close to the surface, and therefore easy to work once water was provided. Some of Victoria's richest alluvial deposits, however, were deep underground and so working them led to a method of mining unique to the colony. Ironically, they were also mines where *too much* water tended to be the problem.

These 'deep lead' mines were located along riverbeds that had been buried beneath the surface by thick basalt flows from ancient volcanic eruptions. Deep lead deposits were typically more than thirty metres below the surface. To get right down to the gold, miners had to blast through layers of hard rock and then dig out the gravels in the old channels. Often the buried rivers still carried water, and so keeping the tunnels dry enough to work was a constant challenge. In some cases, individual shafts pumped millions of litres per day for several years before production could even begin. And, if the workings flooded, it could be disastrous: machinery could be ruined and mines forced to close if the cost of pumping out floodwater was too high. Lives could also be lost. One of Australia's worst mining accidents happened in 1882 when, after the collapse and flooding of a shaft, twenty-two men and boys drowned deep underground at the New Australasian mine near Creswick.[22]

As well as dangerous, mine drainage was controversial. Pumping water out of deep lead and quartz mines was expensive, and the question of how to share the burden and cost when water flowed from one underground mine into the next was a thorny one. As soon as miners started digging deep leads at Ballarat, for example, they found water flooding their shafts. Steam-driven pumps were quickly introduced to lower water levels, but for this to work cooperation among all the mines along a line of reef, or course of deep lead, was essential. Establishing a system to enable such cooperation – not to mention the shared expense – for adjacent claims proved very difficult, with rules and regulations frequently changing. Some mine owners who pumped at their own expense directly benefited those along the line who did not. Others were put out of work for years because they couldn't cooperate with their neighbours. In 1877 the *Drainage of Mines Act* set out how the costs of pumping were to be calculated for each mine. As late as 1891, however, a royal commission on gold mining described the system as inadequate and inoperative, and it was only as mining faded in the years to come that the issue became less acute.[23]

The industrial pressure placed on water went beyond the world of processing washdirt for gold. Beginning in the 1870s, tin miners in south Gippsland explored the creeks and gullies flowing down to Toora and Welshpool, and deposits of alluvial cassiterite (tin) discovered on the Ovens goldfield in 1853 kickstarted a century of tin mining.[24] Sluicing companies began working major deposits of alluvial tin in north-east Tasmania, New South Wales and Queensland around the same time. Sluicing for tin involved the same sort of water infrastructure as gold – races, dams and flumes – and, needless to say, was responsible for producing huge amounts of sludge as well.[25]

The erosion that scattered grains and flakes of gold through alluvial horizons across Victoria also released and deposited many kinds of gemstones in streambeds. Miners working with pans and sluice boxes often uncovered brightly coloured gems in residual gravel: diamonds, sapphires, zircons, topaz, rubies and garnets. Diamonds were found on the Ovens goldfield from the early 1860s in the washdirt from gold and tin deposits, while waterworn sapphires of various colours were dotted across Victoria. The colony's potential as a gem field was recognised in reports by the Department of Mines and various exhibitions in the 1860s. But despite the richness and diversity of gold rush–era stone discoveries, the decline of alluvial mining from the 1880s twinned with the discovery of rich opal and sapphire fields in other parts of Australia meant that interest in Victoria's gem fields gradually died.[26]

Quartz mines also needed good supplies of water, both for steam engines to drive machinery and to process ores and gravels through stamp batteries. Not only that, but the water needed to be clean. Some miners could pump water from shafts deep below ground level and reuse it in their operations, but hard or alkaline water caused scaling and corrosion in boilers, so among these large, often highly capitalised businesses clean water was at a premium. One of the colony's biggest mines was the Port Phillip and Colonial Gold Mining Company, which began mining at Clunes in 1857. The company used a total of eighty stamp-heads to crush both deep lead alluvial gravels and quartz reef ores. Each stamp-head required thirty-six litres of water per minute, which meant, alongside other processes, a total daily water consumption of around 4.5 million litres at peak production. With such demand, the company went to great lengths to secure supply. Their first strategy was to pump water directly from Creswick Creek, which involved the construction of a large dam upstream of the town, supplying water to the mine and to other companies nearby.[27] Port Phillip and Colonial even employed their own devoted water supply manager, one Mr Black. A series of severe floods in the early 1860s, however, damaged the works, and when that was followed by a drought in 1865 the company decided to solve its water problem once and for all.

Flowing a little to the east of Creswick Creek was Birch (or Bullarook) Creek, a stream that had, as yet, been scarcely exploited. Rivett Henry Bland, manager of the Port Phillip and Colonial Company, purchased a steam locomotive that had formerly been used on the Geelong to Ballarat railway line and installed it alongside Birch Creek to drive two powerful pumps. Water flowed thirty metres along a tunnel excavated from the creek to a well that housed the pumps. From there, it was forced through a pipeline that ran up the creek bank and into a timber flume that stood ten metres above ground. From here, the water flowed along a three-kilometre-long race and into a dam; and, finally, it was piped from there to the crushing works. This elaborate scheme cost £4840 and commenced operation in April 1867, meeting the water needs of not only Port Phillip but also some of Clunes's residents.[28] But Birch Creek didn't remain untouched for long. When people began to mine the deep leads north of Creswick in the 1870s and 1880s, the stream became a convenient drain for their sludge. This led to fierce disputes, a series of court cases and ultimately the first attempts by legislators to regulate tailings.

Most of the water consumed by the Port Phillip Company was used to wash crushed ore through its stamp batteries. Other mines used water to power machines. For these purposes, water-wheels were particularly favoured in the mountainous country of Gippsland and north-eastern Victoria, where several hundred wheels were in operation by the 1870s. Steep hills and narrow gorges made road building in this region difficult, and hauling in steam engines was prohibitively expensive. With seemingly abundant water and gravity to drive them, waterwheels were the logical answer: by 1863 there were more than 400 in operation on the Victorian goldfields.[29] While for many migrants these were a common and familiar source of power in their home countries, they were notoriously problematic in Australia. Unreliable rainfall and intermittent flow in local streams and rivers meant that often there simply was not enough water to make them work.

Even in the Victorian high country, waterwheels didn't always live up to their promise. Just below Mount Buller, in the Victorian Alps, it's still possible to see the ruins of a goldfield brought down by the failure of water power. In the narrow valley of the Howqua River at Sheepyard Flat, the faint line of a water race can be seen on the hillside. The race has been cut into the steep slope and runs along the contour for four kilometres back up the valley. Eventually, the Howqua catches up to the head of the race where the two intersect. Susan recorded the site with a group of archaeology students in the late 1990s. Two of the students had the job of walking back along the race with a compass and a long measuring tape. After scrambling for hours over fallen logs and pushing through wattle and blackberry, they eventually created a map of the whole race. When they got to the end, where the race met the river, they saw something unexpected. The race disappeared into a tunnel cut through the solid rock of the cliff. Three metres high, tall enough to stand up in and 110 metres long, the tunnel was cut across the narrow point of a bend in the river. Where it rejoined the river, slots had been cut into the rock to hold a sluicegate in place for diverting water into the tunnel. The discovery of the tunnel precipitated another puzzle: why was it there, and was the work of building it worthwhile?

It soon became clear that both the tunnel and the race had been constructed to supply mines further down the valley, and that all the effort involved was ultimately, painfully, in vain. The water was redirected by the Howqua United Mining Company in 1882 to power an eighteen-metre wheel that towered over the nearby Howqua River flats. It was the toast of the Mansfield district, and became affectionately known as the 'Hanney'. Over 100 years later it is still one of the best-known features of the Howqua field.[30] When sample assays suggested the local ore might yield as much as twenty ounces of gold per ton, expectations were high and the township of Howqua Hills was quickly surveyed. Despite the attempts by several companies over the next two decades, however, the field failed, mainly because the difficult 'refractory' ores, rich in sulfides, could not easily be treated by the methods available at the time.[31]

The challenge was taken on by the Great Rand Company in 1899. They intended to trial alternative methods for treating ore, but they were undone by the lack of water. The company spent £32,500 refurbishing the mine, much of it on constructing the race and tunnel that Susan's students found. The longer race provided the power necessary to drive the company's thirty stamp-heads, but it had, alas, taken too long to build. The moment the company seemed about to succeed, the enterprise collapsed. Work on the race finished in the summer season of 1904–1905, just as the river was drying up at the end of the Federation Drought. The company had spent all its money building the race and had nothing left over for mining. Shareholders were called on for extra capital, but to no avail. The Hanney lingered on in the bush near Howqua until 1916, when it was dismantled for scrap metal to feed the war effort.

The failure of the Hanney aside, waterwheels generally made sense in the high country. On the Jordan goldfield on the upper Goulburn River there were around 130 in use in the 1860s.[32] They were also used in other parts of the colony where there were good water supplies, including Blackwood in central Victoria and Beechworth in the north-east.

One of Victoria's most famous waterwheels, the success of which is quite surprising, was at Castlemaine – a place where rainfall is much lower. Yet, in defiance of its natural environment, the Garfield waterwheel is now one of the highlights of Castlemaine's mining heritage. Susan first visited the striking monument in 1989. Two massive stone walls separated by a narrow channel loom out of the bush and thrust eight metres into the air. Each wall is almost a metre thick, rising in a triangle shape with a broad base and flat narrow top (see picture section). The edge of each is stepped like a pyramid, making them easy to climb – in the manner that early tourists used to climb the pyramids in Egypt. There is very little indication, however, of other human or industrial occupation, and little sign of water nearby. It takes considerable imagination to connect these solid stone towers with a waterwheel.

When Peter went back to look at the Garfield in 2012 he uncovered subtle traces of the water that once drove the wheel. The stone pillars that dominate the site were the supports for a huge wheel. The axle rested on top of the pillars, and the buckets or vanes rotated through the channel between them. Water was delivered to the top of the wheel in an elevated wooden flume that extended almost 250 metres back from the wheel to the water source. None of the flume survives, but up in the scrub behind the stone foundations Peter found the place where the flume joined the race bearing transported water. Tucked away up there is a small gap in the wall of the main race where concrete and brick once held a steel gate in place. When this gate was open, water rushed into the flume and nearly 6000 litres of it passed over the top of the wheel every minute. The wheel drove a stamp battery in a shed next door. Today, the channel behind the gate is completely dry; no water flows there anymore. Its size and quality, however, shows the immense effort that went into the construction of these technologies, hinting at how important they once were. This particular race was a metre deep, up to one-and-a-half metres wide, and excavated into gravel and rock. It was built and maintained by the Victorian Water Supply Department and it holds the secret to the success of waterwheels in the dry hills surrounding Castlemaine.

For years after the discovery of gold in 1851, water shortages around Castlemaine and the Mt Alexander goldfield were a problem. Average rainfall in the area is 600 millimetres per year, falling mostly in winter and spring, but the hilly terrain and small seasonal streams made it difficult to build races and dams on a large scale. The original Castlemaine miners clung to their small claims, generally resisting the amalgamation of leases that enabled miners on other fields to work the ground more effectively. Mt Alexander languished long after miners on most other goldfields had solved their water supply problems and begun sluicing on a major scale. All that began to change in 1874, though, when the Coliban System began delivering water through the ranges to the reservoir at Expedition Pass.

The Coliban System of Waterworks was an ambitious and expensive project to supply Bendigo with clean, reliable water for residents and industry. Beset by delays, problems and politics, the scheme took several decades to complete.[33] Drought in the mid-1860s brought the issue to the fore: the locality was desperate for water. Three formal plans for the scheme were submitted to government, including one by John Henry Reilly, the first miner to divert water at Beechworth. His plan involved an open flume of bitumenised timber to transport water from a reservoir on the Coliban, above Malmsbury.

A waterwheel or turbine at the head of the flume would provide power to a sawmill, which would cut the enormous volumes of timber required.[34] Although Reilly's proposal was soon rejected, for a few years he enjoyed a public profile as an engineer and a water expert. The Coliban System was eventually completed in the 1870s, with a large dam at Malmsbury and a 102-kilometre open channel to carry water north to the diggings. The system was primarily intended to supply Bendigo, but a branch was installed to supply Castlemaine and the surrounding areas as well. This required an expensive, mile-long tunnel through the hills near Elphinstone. George Cooper won the contract, and the tunnel was named after him. Eventually, a network of ten channels was excavated to distribute Coliban water around the Castlemaine diggings. The channel system was cut mostly through rock and gravel, and the length of all the channels combined was more than 160 kilometres (see picture section).

The abundant source of water transformed Castlemaine, and a number of local companies sprang up to take advantage of the secure supply. Many used the water for sluicing, but others saw the potential of waterwheels as a power source for underground mining. And so it was that the water boom saw the construction of waterwheels of various sizes: a wheel six metres in diameter, built by the Renaissance Company in 1879; and one twelve metres in diameter built by the Manchester Reef Waterwheel Company. Larger examples included the eighteen-metre wheel built by the Bendigo and Fryers Company, and the biggest of all, an enormous 21-metre wheel built by the Argus Company.[35] Water power became so popular that even established companies converted existing steam-powered operations to make use of it.

One such converter was the Garfield Company, formed in 1882 to work gold in the quartz reefs along German Gully, north of Chewton. Over seventy men worked the mine, and it was well equipped with an eighteen-head stamp battery, a winding engine and a state-of-the-art Tangye pump imported from Cornwall for drainage. Like most underground mines in Victoria, the whole operation was powered by steam provided by two large boilers and a steam engine, and, after three years of operating that technology, the company had produced 7385 ounces of gold and carved out a prominent position for itself in the affairs of the district. By 1886 it was planning further expansion. It already held twenty hectares of mining lease and had taken over the leases and plant of two neighbouring companies. After years of success with steam engines, this was the moment the company decided to switch to a waterwheel. They built the stone pillars, the flume and the wheel itself, which was twenty-one metres in diameter, as big as the Argus Company wheel nearby (the two were said to be the biggest in the colony, if not Australia). The Garfield was highly effective, and laboured for seventeen years until new owners decommissioned the worn-out wheel in 1904.

Garfield waterwheel, Chewton

Why was the Garfield wheel more successful than the Hanney? It seemed strange to us that a modern, efficiently run company like Garfield that could import the best pumps from Cornwall would return to old-fashioned technology like a waterwheel. When they already had boilers and steam engines installed and functioning effectively for extracting gold from ore, why did they spend money and resources building a wheel? The answer was running costs. Water from the Coliban System cost the company one penny per thousand gallons (4500 litres), or £8, 6s per week for eighteen million litres of water. With a wheel, Garfield also saved the wages of a steam-engine driver, then about £120 per year. But the real savings were in fuel. In Victoria most boilers ran on wood rather than coal, and by the 1880s timber around the goldfields was getting scarce and expensive. Miners were notorious for destroying forests and woodlands, chopping down trees for a sheet of bark, to boil a billy or just for the fun of it.[36] To feed the mines, the forests of central Victoria had largely disappeared; the hills around Castlemaine were almost bare. What we see there today is all regrowth forest from the past century; many trees now have multiple coppice stems growing from what was once a single stump. But by the 1880s, woodcutters had to travel long distances for timber, and even then all they found were saplings as the large trees had already been chopped down. Wood was at a premium. Timber cost five to six shillings per ton in the district, and a large boiler easily burned through 100 tons per week. The Catherine Reef United mining company at Bendigo burned through almost 3000 tons of firewood and 1248 bags of charcoal in 1862, as well as tens of thousands of slabs, props and planks. Big mining operations like the Port Phillip at Clunes and the Koh-i-Noor at Ballarat would contract for delivery of 6000 tons of firewood at a time.[37] A gradual recognition of the need to set aside reserves of timber and protect them for the future was emerging. When the Garfield's manager John Ebbott complained in 1890 that mines were 'starving' for timber he must have been glad of his waterwheel.[38]

In the face of this wood shortage, water was the cheaper source of power, and a means to save on operating costs. The real subsidy in the Garfield case, however, had been in securing the water supply to the wheel. On most goldfields, those wanting to erect a wheel had to secure their own water supply, but Garfield got millions of litres of reliable water at a nominal cost. Most companies had to build a dam and construct a race – or, at the very least, an offtake and long flume from the nearest creek. They therefore had to understand rainfall patterns and the local topography so they could put the dam in the correct spot, and establish the correct gradient for the race. The Great Rand Company at Howqua, for one, had spent thousands of pounds on this process, trying to solve the problems that had got in the way of the Hanney and bankrupting itself in the process. But when it came to Garfield, the government had already done the hard, expensive work of building a dam at Malmsbury, surveying the route and excavating channels and tunnels to carry water through the ranges. All John Ebbott and his men had to do was connect their mine to the existing system. No wonder so many companies around Castlemaine switched on the water.

Yet another way of using water was the large-scale alluvial mining technique introduced to Victoria in the 1890s known as hydraulic pump sluicing (see figure on page 59). This process began with the excavation of a large pit, with another depression nearby in which to deposit tailings. A barge or pontoon was constructed in the pit to support and house engines, pumps and other machinery. Water was then diverted or pumped to the dredge to create a jet of high-pressure water that ran through a nozzle to sluice away the alluvial face. Soil passed through channels or sluices cut in the ground, leading to a sump several metres deep. The gravel pump on the barge sucked the washdirt up and onto elevated sluice boxes, which then passed into a tailings dam. Once the limit of stacked tailings in the dam was reached it became necessary to shift the plant to a new site. The pit was flooded to float the pontoon into a new position.[39] The practice peaked in Victoria in 1906, when there were 102 hydraulic pump sluicing companies at work, processing 6.7 million cubic metres of alluvium per year.[40]

Hydraulic pump sluicing, Eldorado

Cocks Eldorado dredge

Another mining method that was notorious for the way it used and abused water was bucket dredging. Bucket dredges were mobile processing plants for working alluvial gravels in riverbeds. They consisted of a continuous chain of buckets mounted on a frame or ladder attached to a large floating pontoon. The buckets chewed into the bank at the front of the dredge and spat the debris out the back. The pontoon sat in an artificial pond created by the dredge itself. Cables and winches hauled the dredge forward and sideways along the valley, excavating the pond as it went, and using the debris to fill it in at the rear. A couple of dredges survive around Victoria, including one at Porcupine Flat near Maldon, which was still operating as late as 1984.

A much larger and more famous dredge lies at Eldorado, east of Wangaratta (see figure on page 59). When it was built in 1935, the Cocks Eldorado dredge was the biggest in the southern hemisphere.[41] It recovered a massive 70,000 ounces of gold and 1180 tons of tin from the banks of Reedy Creek during its twenty years of operation. Now it sits dormant, rusting in its pond on the edge of town. This monument to water and land exploitation is listed on the Victorian Heritage Register, and history-minded groups ponder how much should be spent on its conservation.

We visited the dredge on a hot spring day when rain was threatening. The dredge looks like an enormous floating shed, with machinery on the pontoon hidden by walls of corrugated iron. The whole plant is 100 metres long and twenty metres wide, with a roof that soars high above the deck. The long conveyor belt of buckets comes out one end and the chute for the tailings sticks out the other. We went up the gangway into the shed to see the rotating screens and tables used to concentrate the gold. Inside, the air was stifling. There is a mass of gears and levers used to direct the bucket chain and the fuse boxes on the electrical plant that drove the works. Wire safety mesh prevents visitors from getting too close, but we could see the ladders connecting the various levels and mountings for the processing equipment. Outside again, we followed the track around the pond and across some of the redeposited tailings. Now the sands and gravels are revegetated with black wattle and other scrub species that favour disturbed ground. Upstream from the dredge pond, the tailings fill a 7.5-kilometre path that winds around the town of Eldorado and back up the valley to where the dredge started its work. Over its twenty years of life the Cocks dredge chewed through about 100 hectares of channel and floodplain.

Dredging was introduced from New Zealand in the late 1890s and immediately stirred up controversy because of the damage the process did to waterways and farmland. In Australia, much rich agricultural land along riverbanks – along with the riverbeds themselves – was destroyed before regulations were established to keep dredges out of the rivers. Floating on their ponds, the dredges appeared relatively innocuous, but each one could cover a lot of ground, moving slowly but inexorably along the valleys. Big ones like the Cocks dredge at Eldorado excavated to depths of thirty metres and processed 1.5 million cubic metres of ground each year, twenty-two million cubic metres across two decades of operation. A contemporary of the Cocks, the famous Tronoh at Harrietville, dug down forty metres below the surface, and had a 356-hectare lease that ran seven kilometres along the Ovens River.[42] At Newstead, the Victoria Gold Dredging Company's plant chewed through 190 hectares of land and built a 1.6-kilometre-long artificial channel to divert the Loddon River so its riverbed could be worked.[43]

Bucket dredges were widespread in the early years of the twentieth century. In 1900 there were already six plants operating, and by 1912 there were more than fifty dredges at work around Victoria.[44] In many cases these were used to rework old ground – the mullocky land where generations of diggers had been before. The creeks and riverbanks were the places where mining started, and only something like a bucket dredge could make it profitable to work again. Those same riverbanks had also received all the sludge and debris from mining upstream, so the dredges reworked tailings too. Near Yackandandah, the Lower Staghorn dredge worked an area where, prior to dredging, the surface was 'several feet above the original, the difference representing the deposits of many years from the varied alluvial workings'.[45] Dredges revived many alluvial fields if the river valley was wide enough and rich enough. Creswick had a dredging boom, as did Castlemaine and Avoca, but the practice was really concentrated in north-eastern Victoria. The Ovens and Buckland valleys were crowded with operations, hosting no less than forty-seven of the state's fifty-six bucket dredges in 1912.[46]

Even though dredges worked in their own ponds, a regular supply of clean water was still required for topping up. The Newstead dredge pumped 100,000 litres per minute into its pond to keep the operation going. Most of the time water supply was straightforward because dredges worked along river flats and water was taken from the stream as needed. During the summer months when the river dried up the company used a bore for its supply.[47] One reason these top-ups were needed was that wastewater in the ponds had to be pumped out to keep the ponds clean enough for the machinery to operate. J.H.W. McGeorge, manager of the Newstead dredge, described the waste as 'thickened water', but what he really meant was sludge. By the 1950s, when McGeorge was supervising the dredge at Newstead, companies were required to use settling dams for their waste. The Newstead dredge was regarded by the industry as a showpiece of good environmental practice. McGeorge claimed that 'In practice it has been found possible to avoid discharging any effluent whatever into the river', as the water from the settling dam was pumped back into the pond as part of the topping up.[48]

For the first few years of bucket dredging, however, no such regulations applied. Companies that obtained leases before 1906 could operate directly in the river channels and on the adjacent banks with no requirement to treat their tailings, much less to restore the land. Landholders around the state were outraged by the destruction of fine farming land, and protest groups sprung up in many districts. The earliest was the Kiewa Valley Anti-Dredging League, which formed in the north-east in 1901. It was quickly joined by the Loddon Anti-Sludge Association in central Victoria, formed in 1906, and by the Ovens River Frontage Anti-Sludge Pollution Association (later the Ovens Anti-Pollution Association), which formed in 1907.[49] A couple of decades later, *The Age* described the emergence of these organised agitators as 'an open rebellion'.[50] We will return to this confrontation later.

Different forms of mining and different types of soils produced different kinds of sludge. Puddlers in Bendigo in the 1850s were working heavy ground with a great deal of clay and silt. They knew the process intimately, and evidence they gave at the 1859 Royal Commission into the Best Method of Removing Sludge from the Gold Fields provides vivid insight into the properties of their detritus. Crawford Mollison, the resident mining warden, described conditions in the various gullies on the field and the different kinds of tailings produced:

> One of the most important things the Commission should take note of is this, that in Bendigo we have two different kinds of sludge. At the White Hills the diggers use long toms [sluice boxes], and run their sludge into the dams they have constructed for that purpose; and no sooner does it get into them than it settles, and the earthy particles immediately separate, and the water is used over and over again. [Elsewhere on the

field] I have seen thousands of paddocks of sludge from the red clay, together with the dam water that was collected; but I have never yet seen any thin film of water overlying the earthy matter, for the water is separated by evaporation.[51]

At the White Hills, the gravelly soils allowed sludge to be trapped in settling dams, but the heavy clays elsewhere caused real problems: their fine soil particles did not settle out, but remained in thick solution for months, even years, until the water evaporated. When the sludge was held in a settling dam it dried from the top down, leaving a desiccated, solid crust above a semi-liquid layer. Puddler Gilbert Browne described the result: 'you can walk on the top of a paddock, and the bottom will remain, at three or four feet under, the same as when it was put in ... I have known it to remain as long as five months, and then the bottom had not become solid.' In Edward Nolan's opinion, such dams were 'injurious to both life and property – it will get hard after a week, so that you can walk on it, but if you take the top scum off, you will find it thin below, and many men have been nearly drowned in it'.[52] As we have already seen, people and animals sometimes *did* drown in sludge.

In contrast to Bendigo's clay, the stamp batteries at Ballarat pumping their waste down the Leigh River towards Geelong produced a mainly fine sand that was the result of crushing quartz. Sluicers created sludge that combined the clay, sand and gravel that had once trapped gold in the beds of ancient rivers. They stacked the boulders and cobbles in long tailings mounds in the base of their sluice pits, and let the finer clay and sand wash out through tail races. Bucket dredges stacked the gravel and sand behind them in ponds, but the finer particles of silt and clay created the 'thickened water' that escaped into rivers.

Once discharged from the mines these different kinds of sludge behaved in different ways. The largest, heaviest portion stayed close to the work site, and indeed remained there in many cases: these are the piles of cobbles and gravel still present at old workings that have provided local councils with a source of road metal ever since. Most of the time this material did not actually get into the rivers at all but remained on the banks and floodplains.

The fine clays and silts were the lightest; diluted sufficiently, they were carried away fastest and travelled the furthest. They immediately washed into rivers and discoloured the water for tens of kilometres downstream, making it undrinkable. It was the silts that carried the heavy metals into the Ramsar-listed wetlands at Reedy Lake near Geelong. Under normal conditions fine particles like this remained in-stream and did not reach the floodplains. If storms caused a flood, or if the river's flow was obstructed by a bridge or a dam, or if mining produced too great a volume of water and sediment for the stream to carry, then the sludge spread out over the banks and onto the flood-plain, as it did north of Bendigo and west of Beechworth.

It has been many years since those fine particles of mining waste were carried in Victoria's rivers, but the layers of new sediment on the floodplains will remain there more or less forever. Ploughing on the flats disturbs them, but leaves them in place. It is only along the riverbanks that the old sludge is sometimes removed. When there is erosion along the river, some of the sludge falls into the water again and is carried further downstream. Sludge on the banks is also released as the river incises a wider bed.

The medium-sized portion of sludge – the sands – has stayed in the rivers the longest. Sand was an in-stream component of sludge that was light enough to be washed into waterways but too heavy to be carried away quickly or to move very far. It settled on the river-bed and stayed there until the next storm. It was sand that filled the deep waterholes on Bet Bet Creek, downstream from Maryborough, and sand that raised the bed of rivers so that bridges had to be rebuilt time and again. Large storms and flooding that made the river flow faster pushed the sand downstream in pulses for short distances at a time. Geographers first noticed these sand slugs in the rivers downstream from California's mines.[53] Large slugs like those can be several metres deep in the riverbed and extend over distances of hundreds or even thousands of metres. Sand slugs in Victorian rivers are, to this day, still carrying mining sludge downstream.

Sludge was carried into rivers by the prodigious quantities of water that miners used each day. From the 3000 litres per minute needed to wash crushed ore through the Port Phillip Company's stamp batteries, to the 100,000 litres per hour pumped into the dredge pond at Newstead, to the 6.7 million litres pouring each day through John Pund's sluices at Beechworth, the mines were Victoria's biggest water users. John Reilly pointed out in 1864 that a small sluicing party of just six men used more water than 1000 families in Melbourne.[54] Miners used more than four times the amount of water consumed by the rest of the population combined. The average daily consumption of water in Melbourne in the 1880s was around fifty gallons per person, or 227 litres.[55] Victorians may have used as much as 200 million litres per day for drinking, washing and cooking shared among the 900,000 people then in the colony. Miners at that time easily provided more than four times that amount, capturing enough water to supply the daily domestic needs of a city of four million people. The enormity of water supply on that scale raises the question – where were they getting it all from? As we began to realise, getting the water in the first place was not as easy as you might think.

3

WATER

OUR INVESTIGATIONS OF MINING HAVE TAKEN US TO abandoned mines all over Australia, from the copper lodes in South Australia's 'Copper Triangle' to the tin sluices of north-eastern Tasmania and the goldmines of New South Wales and Queensland. At nearly all these places we've observed a particular kind of dry ditch, snaking through the bush. Typically these channels are about a metre wide and half a metre deep, with mounds of dirt piled up on each side. In wetter years, after the break of the Millennium Drought, we even found some of these ditches and dams holding water.

We walked along one such ditch at Eureka Reef near Castlemaine with our young children on a hot spring day in 2011, while we were exploring the site of an old quartz mine. As we wandered among trees, past ruined houses and a collapsed mine chimney, we had to cross and re-cross a running stream contained within a carefully engineered channel. This was a short time after the Millenium Drought had broken and the sight of running water in the bush was a novelty to our little boys (see picture section). They splashed about cheerfully, devising their own engineering schemes, while we wondered about the ditch and all the others like it that we had seen over the years. What were they? How did they work? Who built them and why? The answers to these questions turned out to be quite a story.

John Henry Reilly was the first known builder of one of these water races in Victoria. Fresh from the Californian goldfields where he had been involved in building large-scale water supply systems for sluicing, Reilly and his party tapped a tributary of Yackandandah Creek in late April 1853 by building a race almost three kilometres long, which flowed into a reservoir close to Upper Nine Mile Creek.[1] Water accumulated in the dam overnight, and in the morning was released into a wooden sluice box, forty metres in length. Men would stand on either side of the sluice box and shovel dirt into it. The gold washed out by the water was trapped at the lower end of the box by small crossbars placed across the base. When the entire surface of the ground had been dug down to a metre or so, the sluice box was shifted sideways a short distance and the process began again. The average yield for the seven men at work in this fashion was reported to be one ounce of gold per man per day. With the precious yellow metal selling for around seventy shillings an ounce, they were making more than £20 per week – a fortune for working men of the time. Reilly later claimed that he'd recovered the full cost of constructing the race within a week.[2] The race also provided drinking water for miners in the camp.

Reilly's race was built initially to supply his claim on Snake Gully at the new Nine Mile Diggings (later known as Stanley), but what he really wanted was the water itself. He was an engineer by original training, not a miner, and he saw great potential in supplying others with a resource they lacked. So, as soon as he'd finished the race to Snake Gully, he started on another channel to supply diggings at nearby Long Dick's Gully. Long Dick's was rich in gold but not in water, so miners had been hauling their washdirt elsewhere for processing. Reilly saw the market and moved swiftly: he built a race over a mile long that diverted water from Nine Mile Creek to a reservoir at the top of the gully. From there it ran down the length of the gully before re-entering the Nine Mile. A local reporter wrote with admiration that 'cradles are at work along the length of this artificial stream, and the sum of seven shillings per day is gladly paid by the diggers for each cradle'.[3] The water was used and reused multiple times as it flowed down the gully. More than sixty years later, early prospector Charles Grey Bird (writing as 'Alpha') claimed that Reilly made 'virtually 100 guineas weekly' with his monopoly on the local water supply.[4] Reilly himself claimed that he made from £50 to £80 *per day* from the race, an amount far in excess of what the average goldminer extracted from his claim.[5]

Clearly there was money to be made. By the end of May 1853 others had quickly followed Reilly's lead and there were half a dozen companies sluicing around Beechworth. By early July of that year sluicing had become commonplace across the colony.[6]

Reilly and his peers were taking advantage of new regulations that had been gazetted at the beginning of April 1853 – notices of which tended to be printed on weather-resistant cloth and nailed to a prominent tree on each goldfield. For the first time, these regulations permitted cooperative parties to claim larger blocks of land of up to two hectares in size – an area about the size of the Melbourne Cricket Ground, which was a far cry from the eight-feet squares originally allotted to individual claimholders. Miners were also now authorised to carry out 'Sluicing Washing … at running streams'.[7] Although there was no explicit mention of water diversion, those like Reilly interpreted freely and started digging races.

Together these two key industrial and administrative changes – ground sluicing and mining regulations – ushered in a new way of supplying water, and over the next few years hundreds of companies all over Victoria built races to supply their claims and sell water to others. The achievement was remarkable. Entrepreneurs and private capital were able to do what the government could not: reliably provide enormous quantities of water. In a report on Reilly's activities delivered to Melbourne's *The Argus* newspaper, the Beechworth correspondent observed that:

> While the Lieutenant Governor, Executive, Legislative and Municipal Councils, together with all the official staff of scientific officers of the Colony, have been for many months discussing and blundering over plans for supplying Melbourne and Hobsons Bay with fresh water; and while the colonial government engineer has been found utterly unable to devise a plan for forming a canal from the Bay to Melbourne; the enterprise of a private party of diggers ... led the way in the construction of water-works for this colony.[8]

By the late 1860s the races and dams built by miners on the goldfields were able to carry more than *1.1 billion* litres of water every day.[9] Melburnians had fresh water from the reservoir at Yan Yean by then, but many towns in Victoria were still struggling to access municipal supply. Despite funding from the government to build storages on the goldfields, many of the town dams were too small, poorly located and badly designed. *Dicker's Mining Record* called them 'fish-ponds rather than reservoirs'.[10] An inquiry in 1863 that took in Ararat, Smythesdale, Maryborough, Dunolly, Maldon and half a dozen other towns detailed leaking reservoirs and shoddy construction. Ballarat had eventually secured its water from the forested ranges east of town, but at the cost of a protracted conflict with sawmillers and residents of Bullarook Forest, who lost access to water and other resources.[11] The Coliban System for Bendigo was notorious for its delays and cost overruns, and was not completed until 1877.[12] Water supply had been an ongoing thorn in the colony's side.

Prior to the arrival of colonists, Aboriginal people managed the land in a way that maintained waterholes and ensured enough water was available year-round. However, the domesticated animals brought by Europeans both demanded more water and damaged that which already existed by trampling and muddying creeks and waterholes with their hooves. Lack of water has been a problem for European colonists since they arrived on the Australian continent.

In theory Victoria is relatively well served by rainfall. Melbourne's average annual rainfall is about the same as London's – around 600 millimetres. Along the Great Dividing Range in central Victoria, Ballarat and Creswick do better with an average of 750 millimetres per year; and in the high country of the north-east, rainfall can be up to 1500 millimetres per year. North of the Divide things become much drier: the old mining towns of Bendigo, Wedderburn and Inglewood average only 400 to 500 millimetres per year. And, as we all know, rain does not fall in averages. It can arrive with giant storms in winter and spring, or completely vanish in summer. It comes in wet years, with floods that cover the land, and then disappears for years at a time during droughts. There are few natural lakes across Victoria to store surface water, and rates of evaporation under the hot sun are high.

Compounding the unevenness of Victoria's rainfall, its rivers are small even by Australia's generally dry standards. The Murray–Darling, Australia's largest river system, ranks fifteenth in the world by length, but carries only a fraction of the water volume of even modest rivers in wetter countries, and that includes the extra water it now gets from the Snowy Mountains hydroelectricity and irrigation scheme. As the Murray is located on the New South Wales side of the border, the largest river flowing through Victoria is the Goulburn. Based on flow volumes, the Goulburn ranks only forty-third in Australia, well behind much bigger rivers in the tropical north. The rest of Victoria's rivers are even smaller. The Ovens has only half the flow of the Goulburn, and the Loddon and the Avoca are smaller still, often diminishing to a chain of ponds during summer.[13] Only in Gippsland, where the rivers flow south from the Dividing Range into Bass Strait, are the waterways more reliable. Most Victorian 'rivers' are really only little more than creeks. Until gold was discovered there was just enough water for the colony to get by.[14]

The gold rush changed the demand for water – increasing it by several orders of magnitude – just as it changed everything else. To begin with, miners were reliant on the water available in local streams. Diggings were identified either as 'wet', where water was available for working, or 'dry', where washdirt had to be hauled to a water source. Diggings were also seasonal. Ballarat, for example, could generally be worked almost all year round, but Bendigo was well known as a 'winter diggings' that became quiet over the summer as the water dried up.[15] When heavy rain fell there in April 1854 miners built dams out of sandbags to store the water.[16] Beechworth was a wet diggings, with high rainfall, good streams and natural springs, so John Reilly did not have to look far to find a water source for his claim. The first water race there was thus fairly short at less than three kilometres. But on other diggings miners had to work much harder to find the water they needed. On goldfields like Bendigo, filthy water and poor sanitation led to devastating outbreaks of diarrhoea, dysentery, cholera and typhoid.[17] On the other hand, with the potential riches to be had when water was available, it's not surprising that miners were highly motivated to organise better supplies.

We heard about one of the old mining dams near Creswick and went looking for it in the bush one day. We followed a fire track and then cut through the scrub to the creek line. Eatons Dam had once been a popular recreation spot for locals who used it for picnics, birdwatching, shooting parties and fishing. Daryl Lindsay, the writer who grew up exploring the bush around Creswick, described the area in his memoir, *The Leafy Tree*. He and his friends would take their lunch, 'a billy can and sometimes a small swag and camp for the night at Eaton's or Bragg's dams to fish or catch yabbies in the early morning'.[18] The beautiful lake was long gone by the time we visited; the reservoir is now empty and tall eucalypts grow where fishing parties once rowed their boats. Now, the most striking feature of the landscape is the dam wall, a stunning dry-stone structure covered in moss and overhung by grasses and ferns. It's four metres high and seventy metres long, and the top of the wall is broad enough to walk on. We followed it out to the middle of the dam and admired the dry-stone buttresses built to fortify the wall on the downstream side. What we were really interested in, though, was the big gap in the dam where the wall had been breached. The break gave us a perfect view of how the dam had been constructed. It showed us that the masonry wall was only a facing on the downstream side. Behind the facing we could see the earthen embankment that gave the structure its real strength. Like an iceberg, what is normally visible of dam walls is only a fraction of what is concealed by water. In this case the narrow path we had walked along was merely the apex of a structure that was up to eight metres wide at the base.

When the Eatons Dam was built in 1862 the local papers were full of praise for the 'energy, perseverance and liberality' of its construction, and how well this reflected on the local mining community.[19] Seventy years later, when the dam failed spectacularly during a late spring storm in 1933, the papers had changed their tune:

Flood started Wednesday 29th November. Six and a half inches of rain. Waters dissipated quietly after initial flooding ... but by 1 pm [those affected] were disturbed by the ringing of the firebell. Quickly the alarm had spread that Eatons Dam had gone. Hundreds of people hurried down to the creek to watch for the on rush of waters but the alarm was false. Eatons Dam had gone but no one knows exactly when, for reports were most contradictory. Sifting them all through however, it seems that a small piece of the bank must have given way about 6 am and that throughout the day the cut wore deeper. There is ample evidence that at one stage the water was flowing over the embankment for its whole width. One man who went up to investigate reported that at 9 am the bank was still holding but a small hole had appeared in the top and further showers would probably cause the dam to burst. Another person saw the dam on Saturday when it still penned back a huge quantity of water. By Sunday morning however, the cut had eaten down to the bottom of the bank and the creek was flowing [in] the old course.[20]

Fortunately, the dam was positioned far enough back in the Creswick State Forest and there were no homes or businesses downstream to suffer damage, but the picnic spot was destroyed – left to be reclaimed by the bush.

Our cutaway view of how the dam was built offered some clues as to why it had failed (see figure on page 75). On our numerous visits to Eatons over the years, we've brought various colleagues to the site. Hydrologist Leon Bren and geomorphologist Ian Rutherfurd, both experts in the behaviour of water in streams, rivers and dams, observed that the methods used to build Eatons were unusual. Simple earth-fill gravity dams normally consist of a roughly symmetrical mound of clay heaped across a watercourse. Ideally the clay is puddled in successive layers to remove impurities and consolidate the mass, with a central watertight clay core dug down into bedrock to provide a seal.[21] The dam or embankment provides a physical wall

Eatons Dam wall

to resist the vertical and horizontal pressure exerted by the depth of water stored behind it. When Ian and Leon examined the cut where Eatons Dam had failed, they could see that the dam lacked the mass on its downstream side to provide resistance to the weight of water. Instead, the clay was simply banked against the stone wall. The whole structure had only about half the bulk typically found in a gravity dam. The most likely reason for this was expedience and cost reduction: miners needed to create water storage quickly and cheaply, and dams were not, on the whole, intended to last a long time; they were needed only for the life of the mine, a few years at most. The steep stone buttress in the centre of the wall at Eatons may have been intended to provide extra support, but it was not enough. The lack of mass and slope on the downstream side meant that when water poured over the dam during floods it undermined the foundations until eventually the whole thing gave way. The dam's bywash,

a channel at one end of the wall to divert high flows, may also have
been too small – overwhelmed by floodwaters, it might have allowed
a breach to start in the top of the wall.

Eatons Dam was built by two main players on the Creswick goldfield, the brothers Benjamin Franklin Eaton and Charles Lafayette
Eaton, who were born in Ohio in the United States. Benjamin and
Charles began their mining career in Australia on the Turon goldfield
near Bathurst, New South Wales, in 1853, where they were known
as the American Company. Working on a 'gigantic scale', they used
their Californian experience to sluice a stretch of the river with a race
three metres deep, lined with boards and slabs[22]. After six months
they were recovering twenty to thirty ounces per day and employing
thirty hands at weekly wages of £3 per man.[23] By December 1854,
however, their gold yields in Bathurst had declined, and they took
themselves off to Victoria, where, for many years, they would dominate mining and water supply. The Eatons formed and reformed
syndicates with several other big players and fought long-running
disputes with others. The races, dams and other infrastructure they
built controlled the fortunes of hundreds of miners.

In 1857 the Eatons purchased the Yankee Dam and water race on
Creswick Creek for £500, employing up to fifty men at 25 shillings
per week. Yet only two years later, with sluicing at the mercy of dry
conditions, they were declared insolvent. With debts more than twice
their assets, shares in their waterworks were offered for sale. In addition to fickle water supplies, these financial problems may have had
something to do with the invention of a mechanised sluice box, which
used a horse or steam engine to drive sluicing rakes in a trough up to
thirty metres long, and which they patented in 1857. They set one up
at John Kirk's claim at Ballarat in 1857, but the concept never took off.

Somehow, the Eatons mined or traded their way back into the
black, and by 1862 they had begun construction of the large new

storage reservoir on Creswick Creek that became known as Eatons Dam. Once it was finished they dismantled and drained Yankee Dam, washed the creek bed for gold and sluiced the tailings that had accumulated behind the dam from earlier workings upstream.[24]

The next project for the Eaton brothers was a race that transported water from their dam three kilometres downstream to Portuguese Flat. For this they went into partnership with John Roycraft, another big water player in the district whose family was active in mining water supply right into the 1920s. Benjamin Eaton then got involved in maintaining the town water supply while leasing the excess water for his own mining claims. For decades, councils and miners oscillated between competing and cooperating in securing adequate water, each using the other's supply systems to meet their needs.

Meanwhile, Charles Eaton was sharing his considerable thoughts on river snags with the NSW government. With the claim that he was familiar with removing logs and debris from the Mississippi and its tributaries, he offered his expertise to the Select Committee on Murray River Navigation.[25] It appears that by around 1870 the Eatons had moved on from the Creswick mining scene. Benjamin got married a few years later and had a daughter, Maud; he later became a librarian and eventually died on the Mornington Peninsula in 1894, aged seventy-four.[26]

'Water is only another name for gold', *Dicker's Mining Record* declared in 1863. As the periodical rightly pointed out, yields across the colony fluctuated in direct relation to the abundance or deficiency of water.[27] Dams like the Eatons were built on most goldfields in the 1850s and 1860s. They were the necessary first step towards better water supply.

We have come to know the dams around Creswick intimately across the course of our studies. Creswick is a well-documented field with good archival sources and a rich collection of archaeological sites preserved in the Creswick State Forest. It has proved a good location for our research into mining, water and the sludge phenomenon. There isn't much written evidence about mining dams, but the more we visited archaeological sites around Creswick, the more we began to unravel how and why the dams had been constructed. We discovered that they were built by hand, using picks and shovels, wheelbarrows and horse-drawn scoops. In narrow valleys, clay and rock were excavated from adjacent creek banks. On flatter, more open sites, material was dug from the floor of the reservoir and dragged forwards to form the dam wall. The clay was then piled across the stream to form a simple embankment in an inverted 'V' shape. At several sites we visited, you can still see the iron pipe with a valve or stopper that had originally been inserted into the lowest point of the embankment to control the release of water. More commonly, though, there was a large gash in the dam wall down to creek level where the pipe and valve had been dug out for re-use elsewhere.

Miners' dams were clearly intended as temporary structures, built to last for a few years while gold yields were high. Dam walls around Creswick were mostly low, around one to two metres tall – so Eatons was a rarity at four metres in height. Most tended to be the size of the small dams you would find on a farm today. The clay was often piled steeply to get as much height with as little material as possible. That might have been what the Eatons were trying to achieve with their stone wall at Creswick. Simple earthen dams relied on their mass to resist the pressure of impounded water, and less mass meant a weakened structure. Looking at the dams they built, it's clear miners often cut corners, using less clay and neglecting features like the puddled clay walls that would provide an impervious inner barrier. When floods came the dams failed and released a cascade of water down the creek.

Bragg's Dam is one of the better ones. Built in 1859, the wall runs 110 metres across the wide valley of Creswick Creek, upstream from Eatons. It is relatively tall (4.7 metres) and was built at a low angle with a wide base. Around 3500 cubic metres of fill were used, giving the wall sufficient mass to hold back the ninety million litres of water that sat behind the wall – and covered more than 3.6 hectares – when it was full. The dam was not always so bulky though. The first structure was 'substantially built of wood', according to contemporary newspaper reports, and the clay was a subsequent improvement.[28] Today, Bragg's Dam is mostly silted up and holds little water, but several mining dams in the area built at the same time remain part of town water supplies: Russells Dam at Creswick and Kirks Reservoir at Ballarat, for example, were built by miners either side of 1860 and, in their much upgraded forms, are still in use today. Another modern example is the Talbot Reservoir at Evansford, originally built by Stewart and Farnsworth from the late 1850s and purchased by the Borough of Amherst in 1875.[29]

Another prominent force in water development in the area was the Humbug Hill Sluicing Company. It was led by John Boadle Bragg, a naturalised American born in Ireland, who worked as a tanner in New Orleans before arriving in Melbourne on the *Baltimore* via New York in May 1853, at the age of thirty-two. Bragg's partners were a colourful bunch, which included Jacinto de Lima from the Azores Islands in Portugal; Domingo Francisco, a Philippine sailor who had jumped ship in Sydney in 1853; a publican from Kent in England called James Videan; and John Williams, who may have been Cornish. This motley crew soon became part of one of the most ambitious water companies on the Creswick goldfield, building a large storage reservoir, more than twenty kilometres of water race, and lengthy flumes and tunnels. Despite the early death of John Bragg in 1865, Humbug Hill Sluicing remained a going concern for more than twenty years. And they were very combative – in the next chapter we'll return to the long-running series of disputes they had among themselves and with the Eaton brothers over water.

Once we started looking at dams more closely we realised there were different kinds for different purposes. The largest mining dams, like those of the Eaton brothers and Braggs that were built across rivers or streams, stored winter and spring stormwater to ensure supply through dry periods. Smaller holding dams, receiving supply from larger storages, were to hold water close to mining claims. These offered the benefit of slowing water down by spreading it over a wider surface before it was distributed to the working site.[30] They would also accumulate the water that flowed in through the night, so that a supply built up for use during the day. Sluice dams were smaller again, perhaps holding only 500 litres, and built on top of the bank above a sluicing claim. Water entered the dam constantly from a supply race and there was a hose extended from the dam down to a hydraulic sluicing nozzle.[31] Around Castlemaine these were called 'poddy dams' and they created a head of pressure for hydraulic jet elevator sluicing.

Water stored in mining dams was an attractive prospect for everybody in the vicinity, especially in dry conditions. In the Beechworth area, it was tacitly acknowledged that anyone could take as much water as they needed from a race, free of charge, for domestic purposes.[32] Small groups of miners could also apply to reserve a particular spring or waterhole for domestic use. As the government was handing out so many water permits and licences to goldminers, it was accepted that authorities also had a duty to ensure water supplies for public convenience. By-laws in most areas allowed water in mining dams to be used by neighbours and local people in the same way.[33] Jacob Braché, a Prussian mining engineer from Forest Creek, told a government select committee in 1856 that as soon as he had built his dam, which held about 4.5 million litres, 'the people of the surrounding country came down and took the water as they wanted it', especially horse thieves and ex-convicts. Braché had been trying to create crushing machinery, but local miners feared he was establishing a monopoly over both claims and water supply, so they took matters into their own hands and drained the water as soon as it collected.[34]

'Sluice-heads' was the amount that miners used to estimate water volumes, a vague measurement that varied from place to place. A sluice-head referred to the amount of water flowing through an open wooden box of specific dimensions placed at the head of a water race where it joined a creek or dam. The actual amount of water that made it all the way along a race to a mining claim was a different matter. The box had a sliding gauge at midpoint that lowered to restrict the flow to a notional volume over twelve or twenty-four hours, with water backing up behind the gauge serving as a rough measure of pressure. Mining districts and divisions used different sized sluice boxes, which resulted in wildly varied amounts of water. The original Beechworth sluice-head was 36,000 gallons, or 160,000 litres, per twelve-hour day, for example, even though the exact flow rate was often disputed.[35] By the 1860s, however, the measure of sluice-heads was more formally specified in district by-laws. Around Ararat, for example, miners employed a box almost two metres long and thirty centimetres wide with a five-centimetre gap at the bottom, which provided more than ten million litres per day. Miners at Maryborough got a bit less – 8.4 million litres. At Beechworth, however, the gauge only permitted a daily flow of 800,000 litres, while Ballarat miners got about 950,000. Secretary for Mines Robert Brough Smyth acknowledged in 1869 that the sluice-head had long been a subject of controversy;[36] and when British engineer Richard H. Sankey visited from India in 1871 to inspect progress on the Coliban scheme, he concluded that he was 'wholly at a loss' as to what the term sluice-head actually meant.[37]

By the late 1850s, most of the 400 or so puddling mills in the Castlemaine district were owned and worked by Chinese miners, who often used steam to power their supply water.[38] Rules and regulations for using puddlers were translated into Chinese characters and displayed as notices around the goldfields, stating where and how to construct puddlers and spelling out the official powers of field officers and headmen. A printing woodblock from the 1870s prepared by chief Chinese interpreter Charles P. Hodges includes details on how to construct water channels for puddlers and how to maintain sludge channels for drainage, even when the puddler had been decommissioned for some time.

Water is heavy, so it was always easier to let gravity do the work of moving it around. The Staffordshire Reef Quartz Mining Company near Ballarat, for example, built five dams stepped down a nearby gully. Sometimes, though, water supplies could only be captured at a level below the mine and so pumping was required. The Independence Quartz Mining Company at Ballarat cut a tunnel 100 metres from its dam to a pump shaft, where the water was raised twenty metres up to the stamp battery. At Smythesdale, the Nintingbool Sluicing Company stored water from winter rainfall in small dams along a creek, and used a ten-horsepower steam engine to raise the water up the hill to a 100-metre-long flume. From here it was used to sluice a face several metres high at the brow of the hill.[39]

Quartz miners and puddlers also built dams on their claims and sometimes used water pumped up from deep underground shafts. In the early years of the gold rush, few miners at Bendigo had bothered to construct even modest earthworks to trap surface water, but drought taught them the folly of their complacency. By 1871, when Thomas Forbes, a Bendigo mining surveyor, took stock of water supplies in his district, most miners had become more self-sufficient. The Coliban scheme was still years from completion, so Forbes was interested in all of the small, private water storages then used for reef mining in the Bendigo valley. His survey identified 208 reservoirs, with an aggregate capacity of about 730 million litres.[40] Independently, most held less than eight million litres, with several very small dams used for puddling works. The big ones were in Long Gully, a small tributary of Bendigo Creek just west of the town centre. Three of these reservoirs were associated with a large mining tenement on the Perseverance Reef held by Henry Koch. He ran a stamp battery that was one of the largest in Bendigo, using four steam engines and ninety-six stamp-heads to crush 600 tons of quartz per week. Koch also treated pyritic ores from other mines around Victoria. Each of his stamp-heads used 1000 litres of water per hour, which meant that his plant needed almost one million litres per day. Koch also employed Chinese labourers to wash the tailings from his works, demanding even more water. He was reportedly willing to pay hundreds of pounds extra per year for additional water to carry off his waste tailings.[41]

On the Ovens goldfield at Beechworth, men like John Reilly and John Pund didn't need big dams or pumps – they sourced most of their water from creeks and underground springs. They did, however, need races to carry the water to their claims, and in the construction of these transportation infrastructures Pund became another water entrepreneur. In the early days, Pund worked alluvial claims around Yackandandah and Stanley (see figure on page 84). Later he created a water network by buying up existing races and stitching them together. His race system eventually extended almost twenty-eight kilometres, diverting water westward out of Upper Nine Mile Creek, across the headwaters of Spring Creek and into Three Mile Creek. When he started this project in 1865 the area around Three

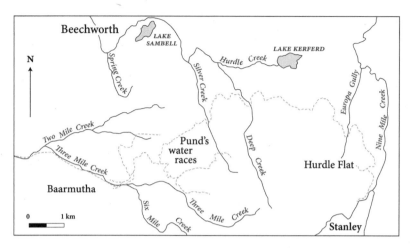

John Pund's water races

Mile had largely been deserted, with only a few Chinese miners remaining. Pund saw the potential. He and three partners applied for a water-right licence – No.58 – and over the next five years built nineteen kilometres of race. Throughout the 1870s and 1880s, Pund consolidated his position by acquiring more mining leases and storage dams. In 1881, for example, his acquisition of water-right licence No.442 delivered an extra 4.3 million litres per day from Nine Mile Creek to the Three Mile sluicing area. These were not trivial amounts of water. By 1887, he held twenty-four hectares under mining leases and miner's rights, and he used hydraulic hoses to sluice alluvial faces up to eight metres high.[42]

Pund was not the only miner to build a water system in the Beechworth area. By 1884 fourteen water-right licences had been issued in Three Mile Creek alone, and the district had more races than anywhere else in Victoria. By 1868, when Pund was just beginning to establish his empire at Three Mile, there were already 1200 kilometres of water races in the Beechworth mining district.[43] At one location, Pund's race was just one of fourteen channels running roughly parallel towards claims further north and west. His race had to cross other races no less than thirty times.[44] In some cases the flow in these parallel races went in opposite directions, as miners battled to divert water to where they needed it.

In the 1890s, Pund expanded again and went into partnership with William Telford, his neighbour, and John Alston Wallace, a businessman with mining interests all over north-eastern Victoria and beyond. Pund, Telford and Wallace amalgamated further water rights to ensure their race network was the major supplier on the Three Mile diggings. They worked their own claims and sold water to miners at nineteen shillings per week. By the turn of the century Pund & Co was averaging 1000 ounces of gold per year.[45]

Major race holders operated in other districts as well, including the Amherst and Talbot diggings near Maryborough, which saw one of the most extensive mining water systems on the Victorian goldfields, constructed in 1871 by Asa G. Farnsworth and James Syme Stewart. The network included eighteen kilometres of head race, almost 140 kilometres of supply races and 240 kilometres of distribution races, along with a further 160 kilometres of abandoned channels.[46] The scale of their derelict races shows how temporary mining ditches were meant to be. Engineer Richard Sankey reported in 1871 that more than 1200 kilometres of races were abandoned or temporarily out of use across the goldfields.

Numerous small races were also constructed around Daylesford. In 1857 Edward Wardle, John Glennon and William Rose began cutting a race that eventually extended thirteen kilometres north from Wombat Creek. When this proved successful they began a second in 1859. It used water from Kangaroo Creek and ran for almost thirty kilometres, including a flume carried twelve metres high over a gully. This 'humble little venture' was so successful it made a profit even after paying rent to the several landholders whose properties it crossed. Water ran reliably through the races for six months of each year, creating employment for up to 150 miners, and returning £60–£80 per week to the race's proprietors. Edward Wardle was later involved with the Coliban scheme and, with Mark Amos, instigated the Fryers Creek Sluicing and Water Supply Company near Castlemaine. The Kangaroo–Wombat Creek race remained in use for almost 100 years and was still providing water for mining in the 1960s.[47]

In the mid-1860s, the potential to apply water to the diggings around Heathcote attracted the attention of John Reilly, who planned to convey water from the upper reaches of McIvor and Wild Duck creeks near Tooborac towards the diggings at Heathcote, and to promote the McIvor Hydraulic Sluicing Company at the same time. It was an expensive proposition. Reilly tried leveraging £5000 out of the borough council to fund the project, and, in return, offered to guarantee 130,000 litres of water per day for town residents. By May 1867 the company had cut twenty kilometres of race and almost completed a large dam of ninety million litres capacity, and yet the works became neglected, and the scheme collapsed. Reilly withdrew from mining and engineering, and appears to have taken solace in drink. He was later judged harshly, his 'fatal conviviality and carelessness' blamed for the failure of his projects. Alpha claimed in his memoirs that the water pioneer had squandered his money 'in wine and women' and became a miserable wreck.[48]

The work of constructing races was enormous. Each channel was dug with pick and shovel through clay, gravel and rock. They were excavated at a slight fall across the contour to ensure a steady flow of water. A fall of about one in 1000 was a good rule of thumb, allowing water to flow at roughly walking pace and limiting the deposit of fine silts in the channel. Surveying the line for the race took considerable bushcraft. The gold rush attracted many skilled men, so qualified engineers and surveyors were available on most goldfields. In other cases, miners would have used traditional methods such as an A-frame with a simple pendulum at the apex and a few marks on the crossbar to check levels, while a spirit level, water in a beer bottle or simple line of sight would do if nothing else was available.

The writing of James William Robertson offers a glimpse of the process of race building. In the mid-1850s Robertson wrote a series of letters to his father back home in New Brunswick, Canada, describing the various mining and water-supply enterprises he was involved in around Creswick. Beginning in 1855, Robertson and his partners, known as the 'Yankee' party, spent several years developing a lengthy race network, eventually extending for twenty-two kilometres from the Bullarook Forest to the slopes of Humbug Hill (see picture section). To construct the race, Robertson's group employed up to forty labourers at a cost of £50 per mile. At the race's terminus on Lincoln Hill the men built a timber flume almost 250 metres long to span a saddle across to Humbug Hill that delivered the water for sluicing. Robertson had worked in the timber industry in New Brunswick and used flumes to transport logs, so he knew how to ensure the flume across to Humbug was robust and functional. 'Races that could be bought six months ago for fifty pounds have been sold as high as five hundred pounds', he explained to his father in 1857, 'I have managed with two others to secure the best water privilege in this district. [It] runs all the year and is high enough to take to Ballarat.'[49] The upper section of the race was completed the following year, 'cut mostly by Chinamen they are satisfied with 6 or 7s [shillings] per day and very steady men to work'. Robertson expected to employ 100 men for sluicing his claim and pay them up to ten shillings per day, 'with the power to discharge any man at any time'.[50]

The Canadian was in the middle of a long career as a successful entrepreneur – only part of it in Victoria. The grandson of Scottish immigrants to Canada, and the eldest of seven children, Robertson was born in December 1823 near the port of St. John in New Brunswick. He worked in the timber industry and on the family farm as a young man, but when the provincial economy fell into recession in the late 1840s he sought a career in the merchant marines. In 1852, Robertson joined a group of eighty-eight people in New Brunswick who purchased and fitted out the newly built brig *Australia* for a voyage to the goldfields. Most were young men and women, or families with young children, and many were skilled carpenters, farmers and mill workers. Historian Clare Wright points out that such cooperative ventures were a common practice at the time, especially among Scots who believed the rights and privileges of prosperity had to be earned to be enjoyed.[51] The *Australia* was owned by its passengers, effectively as a joint stock company, and the crew, just as keen to get to the goldfields, worked their passage out. The ship departed from St. John in August 1852 and arrived in Cape Town in November. The passengers purchased a cargo of gin and reached Melbourne a month later, where they sold the gin, the ship, and its stores and fittings before going their separate ways, although many kept in touch over the following years.[52] Robertson pursued opportunities on the Turon goldfield in New South Wales and around Melbourne, Adelaide, Bendigo and Beechworth before arriving in Creswick.[53]

The 'Canada race' built by Robertson and his partners at Creswick was an ambitious and successful venture. Like Robertson's earlier participation in the *Australia* project, it reflects his capacity to value-add and make a profit from meeting his own immediate needs. The 'Yankee' party's race was the longest in the Creswick district at the time, and Robertson was fully conscious of the value of what he had created. Most immediately, his race provided water for his claim and for those of other miners, but Robertson saw its potential to provide reliable domestic supply as well: the fast-growing town of Ballarat was about twenty kilometres away, and his race was high enough in elevation.

His water never did reach Ballarat, but its potential for domestic use was realised in 1864 when the partners sold the race's upper portion to the Creswick council, which included it in the town's supply system.

Robertson's race also appears to have supplied clean water to Quinn's Brewery at Cabbage Tree Flat, below Humbug Hill. The *Creswick Advertiser* described the brewing complex as a three-storey building with malt-house, kiln, boiler and chimney, the whole measuring fifteen metres by five metres.[54] In a prize-winning essay for the Victorian government in 1860, engineer Charles Mayes expressed admiration for the abundant water flowing into the cellar of the brewery 'to encircle the fermenting and ripening tuns [barrels]' as a means of improving the quality of beer production.[55] This was an era before refrigeration, and beer was mostly brewed close to where it was consumed. A 'constant supply of pure water' was brought down from the Bullarook Forest, which must have been Robertson's race.

By this stage Robertson was in his mid-thirties – the prime of life – but he resisted returning to Canada despite his family's pleas. 'I have been too long here to think of farming in N[ew] B[runswick,] where it takes six months to get food [and] for the other six you want no barns full of hay ... [Here in Victoria] the cattle for our winter is like your August and our gardens full of vegetables all the year', he wrote to his father.[56] By 1861, however, Robertson's interest in gold and waterworks was fading. Life on the goldfields was not all adventure and profit, and he'd endured a series of injuries, illnesses and financial hardships. He worked for a while in the Bullarook Forest timber industry, but news of the gold rush to Otago in New Zealand was enough for him to reconsider his plans. In 1863 he sold his interests in water races to associates for £200 and crossed the Tasman. In Queenstown, he set up a new firm, J.W. Robertson & Co., which prospered supplying Otago goldminers with flour and sawn timber and carrying their supplies in steamships. 'Daddy Robertson' became the first mayor of Queenstown, and was subsequently elected to the provincial assembly. He died in 1876 at the age of fifty-two.[57]

Even the most straightforward races posed engineering challenges. Stone revetments were sometimes used to support races on the lower sides of steep slopes or to repair breaches. Where the race had to cross a valley, either flumes or siphons were needed. These flumes could be quite magnificent structures. One example, the Lightning Creek flume at Mitta Mitta, built in 1886 by the Mammoth Hydraulic Sluicing and Gold Mining Co., was over 200 metres long with trestles up to thirty-five metres high (see figure on page 91).[58] Bushfires have mostly destroyed old timber flumes, but near Castlemaine we have examined the remains of several large siphons – pipes that carry water across an obstacle, a technology that has been used since at least Roman times. An inverted siphon runs down one side of a valley and up the other in a 'U' shape; because water finds a constant level under the force of gravity, the water drawn down a pipe flows up to the same height on the other side.

The siphons along the race built originally by the Fryers Creek Sluicing Company in the forest south of Castlemaine are spectacular examples of bush engineering. One example we recorded at Salter's Creek is a galvanised pipe, forty centimetres in diameter and 185 metres long (see picture section). The pipe was originally constructed from a series of sections each 1.8 metres in length. These were bolted together with lugs and sealed with hessian and tar. Some of the tar coating is still preserved on the outer surface. Each end of the siphon is bedded in massive abutments built of rough granite stonework and connected with the open channel of the race. From the abutment on the north side of the valley the pipe extends down the hill. At the bottom of the gully, three sets of timber legs make a trestle five metres high that carries the span across the creek. On the far side, the pipe climbs up nearly twenty metres to the same elevation on the opposite

The Flume on the Buckland, *engraving, 1866*

hill where it rests in another stone abutment. The siphon is a remark-able industrial relic, hidden away in the bush, and it is one of the best preserved examples of its kind in Australia.

We've visited four other siphons along the same race. There are more that have now become inaccessible, but their location is visible well in LiDAR imagery. In LiDAR views, the cut of the race shows up clearly as a gouge winding its way along the hillside, the only human-made feature visible in much of the landscape of the state forest. The hills are bare and gently rounded, falling to steep narrow gullies that run into the valley of the Loddon River. Beginning at Red Hill, where the sluicing was carried out, the path of the race can be followed on LiDAR all the way back to where it joins the Loddon twenty-five kilo-metres up the valley. At several points, where gullies reached back from the river, the race suddenly stops on one side only to reappear on the other. This is where the siphons were built to convey water down the gully and up the other side.

A syndicate of miners including Arthur Bradfield began reconditioning the Fryers Creek race in the 1930s, using a grant from the Mines Department. Their work included construction of fourteen siphons, with a water-right licence permitting the diversion of thirteen million litres from the Loddon each day.[59] The Second World War interrupted work, but Arthur Bradfield's son Ray continued using the race to supply his sluicing operations for several decades. When Ray's water licence was officially declared void in 1963 it marked nearly a century of water supply, one of the longest usages of a water network in Victoria.

Bradfield siphons were a particular success because they used strong metal pipes made by a foundry in Castlemaine. Some of the earlier Creswick siphons literally came unstuck when the innovative technology they were made from – paper and bitumen – failed. Bitumen was all the rage in the early 1860s as a new material ripe with potential. Using native asphalts, wood and coal tars and, later, petroleum by-products, the manufacture of bitumen products was expanding rapidly in Europe and the United States, and people eagerly sought ways to make use of its elasticity, adhesiveness, durability, water resistance and cheapness. Bitumen paper pipes, made by drawing a long roll of paper through a vat of molten bitumen and then coiling the paper around a cylinder of the required pipe diameter, were one such development. A Melbourne company began local manufacture of bitumen-coated paper pipes in 1860 and found a ready market: municipal councils and mining companies were desperate for pipes that were cheaper and lighter than traditional cast iron, which was mostly imported from England.[60] John Boadle Bragg and his partners in the Humbug Hill Sluicing Company saw bitumen pipes as a means to carry water across Slaty Creek and thus extend water distribution further to the west. In 1862 they commissioned the Patent Bitumenized Pipe Company in Melbourne to manufacture and install almost 700 metres of pipe in an inverted siphon, at a cost of £650. Around fifty people, including the local press, were on hand to observe the water being turned on for the first time. A gun was fired to announce success and the occasion was hailed as 'the greatest practical test that paper pipes have been put to on any of the gold-fields'.[61] Nearby, the St. George's Sluicing Company also used bitumen pipes to replace its timber flume at a cost of £500, and water flowed there on 15 September 1862, only days after the Humbug siphon was turned on.[62]

Alas, the bitumen experiment was not a success. The pipes in the Humbug Hill siphon burst after only two weeks of use. Other users also found that the joints leaked, the bitumen peeled off and the weakened pipes collapsed. One dissatisfied company at Ararat said they would willingly give the pipes away to anyone who would take the 'much vaunted and highly-praised-up paper swindles'.[63] The Humbug Hill partners reported that they 'suffered very considerable loss of both money and time' and replaced the bitumen pipes with cast iron only two years later.[64] St. George's found them 'a source of annoyance and loss' and replaced theirs too.[65]

These curious failures often make up a good deal of the archaeological record because people leave objects, materials and equipment and walk away, abandoning the evidence in the bush, to be found many years later. When we went looking for the Humbug Hill Sluicing Company's siphon we found no evidence of the iron pipes because they were valuable and had been salvaged long ago. What we did find, however, were traces of the melted bitumen. A tiny patch was visible on the bush track that crossed the line of the old siphon. More of it must lie in the bush, because we could see a faint, narrow gap in the forest canopy where vegetation has not regrown. When the company put in the new iron pipes in 1864 they probably abandoned the now worthless bitumen ones where they lay. Bushfires burned the flammable material and melted the bitumen, leaving it in puddles where it still obstructs the regenerating forest.

Tunnels were another way to move water around hilly terrain, but they were difficult and expensive to build. The Great Rand Company used one on the Howqua River to supply its waterwheel. The Fryers Creek Sluicing Company also built a tunnel of over 130 metres in length. It was part of an ambitious scheme by Edward Wardle and Thomas Amos to bring water in a long race more than twenty kilometres from the Loddon River to the Fryers Creek goldfield. Beginning in 1865, a gang of miners drove the tunnel, while others built the flumes and excavated the race. All the work was done manually with pick, shovel, hammer and drill. The whole system, which included eleven timber flumes, took only six months to build but cost £5000.[66] A tunnel at Devil's Gully was needed because too much height had been lost excavating the race and it ended up twelve metres below the saddle instead of crossing over the top.[67] The tunnel mouth is only about a metre wide and a little higher, roughly cut into the angled bedrock. No light is visible from end to end as the tunnel curves a little – an uneven section halfway through shows the tunnel was dug from both ends at once. At the southern entrance the race emerges into a deep cutting that extends almost back to a supply dam. Devil's Gully Tunnel is one of a series of tunnels linking water supply races between the Fryerstown mining district and the Coliban System, the others built as part of the government scheme in the 1870s.

The race and tunnel at Devil's Gully are a puzzle, because on paper the water seems to flow uphill from south to north. Our colleague Jodi Turnbull studied the system for a long time to figure out how it worked. It was only after we walked the race and Jodi analysed the LiDAR tiles that the answers emerged. The whole race falls about 1.52 metres per kilometre, but as it approaches the tunnel it flattens out and becomes level. This meant that when water flowed north from the Loddon race it could be pushed through and go on to supply Red Hill, as originally intended. Later, when the Coliban scheme was finally extended to the area and the systems were linked up, water from the Coliban channel could be pushed south through the tunnel to Bald Hill. At one stage Coliban water was even diverted south through a pipe laid inside the tunnel, at the same time as the Loddon water flowed the opposite way via the very same tunnel.

Miners also used their tunnelling skills to divert creeks and rivers out of their channels and expose the riverbed for sluicing (see picture section). At a loop on the Yarra River near Warrandyte, the Evelyn Tunnel Gold Mining Company used timber piles and 1000 sandbags to create a dam to hold back the river. Meanwhile, a long tunnel was excavated through a rocky spur. The tunnel was a substantial undertaking – 200 metres long, six metres wide and five metres deep – but it only took three and a half months to complete at a cost of £2100. When all of this was ready in August 1870, the river was diverted through the tunnel and sluicing could begin. Despite all the effort, though, high construction costs and poor gold returns meant the company wound up only twelve months later. The dam gradually decayed, but water still flows through the tunnel at Pound Bend and has drawn visitors ever since.[68]

Successful water networks generated considerable wealth for their owners and created a new class of capitalists, the water bosses. 'Water squatter' and 'water boss' were terms used in the Beechworth district to describe men like Reilly and Pund, who created monopolies over water, and distributed or withheld it to suit their needs.[69] The terms described anyone with the capital to invest in water infrastructure, and who could then onsell the precious commodity to others for a profit. It was not a small man's game: building dams, channels, flumes, siphons and tunnels was expensive business, and investors spent thousands of pounds.

One of the biggest water bosses was John Alston Wallace, who joined Pund's scheme at Three Mile in the 1880s and 1890s. Born in 1824 near Glasgow, Wallace was a major mining entrepreneur who dominated the goldfields of north-east Victoria for almost half a century. A tremendously energetic and often controversial figure, Wallace invested in and developed mines at Stanley, Beechworth, Bright, Chiltern, Rutherglen, Bethanga, Harrietville, Myrtleford and beyond, and represented the North-Eastern Province in the Legislative Council from 1871 until his death in 1901. In the 1880s and 1890s he pioneered steam-powered pump-sluicing at Yackandandah and Woolshed.[70] Optimistic and opportunistic, Wallace was always ready to invest in mining and other ventures that he believed in, and by the turn of the century he employed around 1500 people across various ventures in the north-east.[71] He also had extensive interests in sluicing and water. Wallace was outspoken in his support for privatising water, believing that too much of it was wasted on the goldfields, and that water secured by races and dams should be regarded as private property. He argued that race owners should be entitled to sell excess water and exercise the power to decide who gets it.[72]

William Telford was another water boss involved in the Wallace and Pund syndicate. Born in Stirling, Scotland, c.1825, Telford was one of many Scottish emigrants who achieved success on the Ovens goldfield. He had his own claim upstream of Pund's before they amalgamated and he was chairman of the Rocky Mountain Extended Gold Sluicing Company. This company was responsible for a remarkable piece of hydraulic engineering, Lake Sambell, originally formed via the removal of tailings from a massive sluicing pit that is now a popular camping spot in the centre of Beechworth. In the 1870s, contractors blasted an 800-metre tunnel through granite deep beneath Beechworth and drained the tailings into the gorge of Spring Creek below the town. The tunnel floor was lined with sluice boxes to capture the fine gold particles that escaped the sluice works at the diggings.[73] By 1884, Rocky Mountain held eleven licences that entitled them to the flow of more than 22.5 million litres per day.[74] By the end of the century the company had paid out £47,000 for water rights, in addition to the cost of building and maintaining its races.[75]

John Wallace's younger brother Peter was another big water boss. Peter Wallace arrived in Victoria in 1848 and initially worked on a pastoral station in the Wimmera. When John arrived in 1852 the Glaswegian brothers went into business with a store at Stanley. One of their partners was Hopton Nolan, the farmer who was to suffer so much in later years from their sludge at Tarrawingee.[76] The Wallace business grew to include several hotels and the supply of meat to the goldfields. When Nolan turned to farming in the early 1860s the Wallace brothers began developing interests in mining and water. They realised, of course, that controlling large water volumes gave them a key tradeable commodity. Peter moved to Myrtleford in the early 1860s and soon secured two of the largest water rights issued in Victoria. The first entitled him to withdraw the enormous volume of 160 million litres per day from Happy Valley Creek near Myrtleford on the Ovens River, which appears to have driven a large waterwheel and crushing plant for the Reform reef mine.[77] His brother John believed that by drilling 'deep tunnels into the ranges' the flow of the Ovens River could be increased almost indefinitely; he secured a licence to extract fifty million litres per day from the Buckland River.[78] In 1874, Peter applied successfully for a further licence for another 160 million litres from the Ovens, and so, with two pieces of paper, had secured a daily entitlement of more than 320 million litres, enough today to meet the domestic needs of a city of more than a million people.[79] Between them, the Wallace brothers controlled more water than anyone in Victoria.

Water bosses were invariably white men. We only know of two women who held water-right licences in their own names. One was Catherine Miller of Sailor's Creek, near Daylesford. Her fifteen-year licence was issued in September 1870 and entitled her to extract 1.8 million litres per day. Based on her allotted 'gathering ground' of about two hectares, her race was likely to have been a bit less than three kilometres long.[80] Beyond these scant details we unfortunately know almost nothing more about her, except that she soon sold a half share in her race for eighty pounds to miner Pietro Foletti, perhaps so he could work the claim on her behalf. Many women did indeed hold miner's rights, and could have used those provisions to dig races for mining purposes. Because the only surviving documentation is from the licensing system, those who held claims rather than leases have disappeared from written records.

Chinese miners also used miner's rights for the most part rather than licences to gain access to water. Despite the extensive involvement of Chinese workers in alluvial mining, relatively few Chinese water bosses emerged. It appears Chinese miners preferred to lease water from others rather than build infrastructure to control the supply themselves, or they used smaller miners' races. Wealthy Chinese merchants could certainly have accessed the capital needed to construct water systems, but they seem to have preferred to invest their money elsewhere. In some cases, merchants such as Lowe Kong Meng invested in quartz and deep lead mines, especially in the Ballarat district[81], but in general Chinese men on the goldfields worked largely as contract labourers to excavate dams and races, and leased sluice-heads of water from European holders of water-right licenses to work their alluvial claims. Later they took up rights to pre-existing mining water systems. In some cases, they adapted water races used by miners for the additional function of watering their market gardens.

That said, there were individual Chinese who grasped the opportunities of large-scale sluicing on the alluvial goldfields, especially around Creswick. Ah Tan was the first recorded Chinese miner in Victoria to apply for a water-right licence when he took out licence No.38 in 1865 to construct a reservoir and race eight kilometres long at Creswick. Ten years later he applied successfully for another to excavate an eight-kilometre race from Bungaree to Mason's Gully, and build a reservoir of 2.2 million litres capacity. This may have been the same Ah Tan who died aged fifty from tuberculosis in 1880, having spent the last seventeen years of his life living with his white partner, Creswick woman Henrietta Ann Russell.[82]

Ah Fee was another who perceived the market value of controlling water. Also known as Wai Jung Chin, he was a prominent member of the Chinese mining community in Creswick. He and his party of seventy men were busy sluicing thirteen hectares of old ground at Portuguese Flat in 1862, near the junction of Slaty Creek and Creswick Creek. Two years later he had another, smaller, party at work sluicing almost 1.6 hectares at Humbug Hill. He was a storekeeper in Creswick's Chinese village at Black Lead, and was frequently caught up in legal hostilities over encroachment and forfeiture of claims, and disputes over water and livestock. Ah Fee also leased surplus water from the Creswick council in 1880, before dying in 1883.[83] In later years, Ah Soon, Wun Yee and Han Kee each held water-right licences at Creswick as well. Ah Soon continued to lease surplus water from the municipal supply in the 1880s, paying only £40 over three years for up to eight sluice-heads per week. He sublet the water to other Chinese parties and to Christopher Russell at Cabbage Tree Flat for sixteen shillings per week.[84] Ah Soon was making easy money from lots of water.

To keep water flowing and to ensure that miners leasing water received the share they were paying for, races needed maintenance. Water merchants often employed race keepers to do the job. Edward Wardle's party at Daylesford paid a man £3 per week to maintain almost thirty kilometres of race. The McIvor Sluicing Company paid two men and two boys to look after a similar length of channel in the 1880s.[85] Len Goldsmith was a teenager during the Depression and worked sluicing claims around Creswick with his uncle. Young Len was the labourer in the team and it was his job to let the night's water out of the dam and clean out the race. 'It was hard work!' he reported.

> If it wasn't raining there would be no water in the race at all and it would take four and a half hours to get to us once it was turned on ... I would be the fool who had to follow the water. I had a pitchfork to scoop out all the sticks and branches. God you were going, seven miles of race to clean out ... If you didn't keep up, the rubbish would dam up and the race would overflow ... I used to walk across the flumes. They had to be cleaned and checked as well.[86]

William Crawford Walker had the job of looking after the 'Big Ditch' at Granite Flat, a few miles above Mitta Mitta, in the early 1860s. His diary provides an insight into how his days were filled: checking the race for leaks; clearing out fallen branches and other muck; adjusting the height of the head dam with logs and brush to ensure enough water flowed down to miners known as Big John, Jack and the Sailor, and Chinese miners including Ah Yet and Ah Poon. Some of them grumbled that Walker placed the outlet gauges too high, limiting their claim's flow. Walker sometimes patrolled the race at night to prevent theft, and every week or so he would chase up water payments from miners, often with little to show for it. Effective race management was a full-time job.[87]

To make a profit, the water supplied by the bosses had to be reliable, and the gold worked by the miners had to be rich enough to meet the cost. If both these conditions weren't met, a race would fail. Examples abound of companies collapsing because they failed to deliver enough water. One of the most glaring cases was the Lal Lal Waterworks Association (LLWA), formed by Archibald Fisken in the 1860s. Fisken was a wealthy Ballarat pastoralist who saw a way to capitalise on the need for water on the Moorabool goldfield. The LLWA applied for the very first water-right licence issued in Victoria, which provided a virtual monopoly over surface water in a drainage area covering almost 6400 hectares.[88] The scheme involved an extensive network of dams, races, tunnels, cuttings and flumes that would capture water from Lal Lal Creek and distribute it to miners further downstream on the Moorabool around Dolly's Creek. One of the major features was a large reservoir at the head of the system, covering more than forty hectares and holding about 400 million litres of water, making it one of the largest privately constructed mining dams in Victoria. The water was to be distributed through a network of races extending 100 kilometres over the goldfield. By 1863 the scheme was more or less complete. Miners paid £3 per week for each sluice-head of water, which in the Ballarat mining district officially provided nearly one million litres of water per day.[89]

After spending almost £9000 on construction, the LLWA was soon in trouble. The Ballarat Water Commission illegally diverted much of the flow from the headwaters of Lal Lal Creek to supply residents of Ballarat, substantially reducing the amount of water available to the LLWA and creating a legal dispute that took years to resolve.[90] Evaporation and seepage rates were also high, and the low head and narrow races limited the amount of water that could be delivered to miners a long way downstream. In 1872 Fisken himself drained half the water out of the main supply dam that lay near his homestead. He claimed to be concerned about the health of his family, but it was more likely a response to a dispute with his LLWA partners and concern with declining profits.[91] By 1881 the races were overgrown with grass, weeds and silt, and there were few miners left willing to buy the limited water still available.[92]

The LLWA failed in its ambition, with the sheer scale of the enterprise working against its success. Good storage capacity was undermined by losses along the long route to the goldfields. Upkeep of races and flumes was a continual cost, while the scarcity of gold along the Moorabool meant there was never enough wealth to justify the heavy infrastructure investment. The partners saw value in supplying water to Dolly's Creek, but as pastoralists and engineers they were unable to recognise the poor man's diggings for what it was. Unlike many others, the LLWA network was never incorporated into a municipal supply system, a clear sign that it was unable to deliver the large volumes its promoters had hoped for.[93]

Wealth and its opposite, failed enterprise, were one set of outcomes of the water races. Another was the creation of artificial drainage patterns and the massive diversion of water by people from one catchment to others. Water in Victoria no longer simply flowed downstream through creeks and gullies as surface and groundwater. Instead, it was extracted and diverted sideways, channelled to flow on contours, at right angles to its natural drainage. At Creswick, the networks built by Robertson, Bragg, the Eatons, Russells, Roycraft and others accelerated the natural process. In those places, miners collected water at the top of creek catchments and moved it quickly down to where they wanted it. They bypassed the natural drainage network but kept the water within the catchments of Creswick and Slaty Creek.

Other race networks made more dramatic, far-reaching interventions into natural water patterns. Much of the water John Pund used at Beechworth was taken from the Upper Nine Mile Creek at Stanley. The Nine Mile is a tributary of the Kiewa River. The Kiewa flows north before reaching the Murray just below Lake Hume. Pund's races diverted and discharged large volumes of water destined for the Kiewa into Hodgson Creek, a tributary of the Ovens. All the millions of litres of water that Pund and his colleagues used was water moved from one catchment to another. And thus, all the sludge that oozed over the flats at Tarrawingee was carried by water that should not have really been there in the first place.

The same took place around Castlemaine. The whole point of the Coliban System was to make water from the Coliban River available for mining and domestic purposes, but once people finished using the water, it didn't flow back into the Coliban but was discharged into local waterways, and flowed into Forest Creek at Chewton, Barkers Creek and Campbells Creek at Castlemaine, and Fryers Creek at Vaughan. Those streams are all tributaries of the Loddon. So ultimately, the Coliban System used water from the Coliban River to take Castlemaine sludge down the Loddon to Eddington and the Laanecoorie Weir.

How could someone like John Reilly make so much money selling something that nature apparently provided so freely and generously to everyone? How was it possible that people were permitted to move so much water so readily from one catchment to another? Now that we knew where the water came from, how it travelled and where it went, this was the next piece of the sludge puzzle we sought to put in place: what gave Reilly and all the others the right to sell water in the first place?

4

FIST FIGHTS AND WATER RIGHTS

THE MEN WERE FIXING FOR A FIGHT. IT'S NOT CLEAR who struck first – Julius Eggers and Tack Long both later claimed to have been the victim. In any case, they were quickly swinging at each other with fists and shovels. Shouts and cries echoed up the Three Mile valley, and within moments a full-scale riot had erupted among the Chinese and German miners working there. Later reports suggested up to fifty men were involved, though only nine were named in the subsequent court case. Young Gustave Eggers, Julius's brother, was attacked by Ah Pow and Ah Ping, who knocked him down and tried to bury him alive in sludge. Henry Kerstein punched every Chinese face he could reach until he got hit on the head by Tack Long's shovel. One big Chinese miner swung a twelve-foot pole and no one could get near him. Others on both sides wielded sluicing forks, bamboo poles and an axe. Blood flowed from cuts to faces, heads and arms. The outnumbered Germans gradually retreated until one of their mates appeared with a gun and declared the fracas over.

They weren't the only ones fighting. The following year, 1866, saw further violence on Three Mile Creek. Frederick Warr had employed Ah Moon and another man to work part of his claim, but his employees soon began using Warr's water for their own purposes. Ah Moon and his friend claimed they paid for it, and so they could do what they liked with it. When the disagreement was brought to goldfields officials, the mining warden dismissed the case, leaving both sides unhappy.[1] The dispute flared into violence a few weeks later, when Warr and Henry Boyd tussled with several Chinese for control of a sluice gate. Warr allegedly threatened He King and Ah Nam and then shot at them with a pistol. They fought back with shovels. He King and Ah Nam accused Warr and Boyd with assault, but the court dismissed the case, unwilling to find for Chinese against European miners.[2]

Chinese miners had good reason to think the legal system was stacked against them. In 1872 there was a nasty dispute over mining water and races at Two Mile Creek. Ah Tang was employed by the Alma and Red Hill Mining Company to 'mind the water' that ran from a large dam at Six Mile and past his hut. James McCormish suspected Ah Tang of diverting the water for his own use. Despite Ah Tang's denial, McCormish knocked him down and kicked him senseless, hospitalising the water minder for months with a ruptured urethra. The Police Court decided that McCormish had somehow been provoked and let him off with a small fine for this brutal assault.[3]

Scuffles like those at the Two and Three Mile diggings were occurring on other goldfields as well. A few years earlier, on a cold winter's day in July 1860, a similar scene played out beside a muddy ditch in the forest near Creswick. A group of men shouted and shoved, stabbing their fingers at the ground and at the water in the channel. An ugly fight was brewing. Jacinto de Lima grabbed a shovel and swung its blade at John Williams, then knocked Williams's head against the stump of a tree. Williams struggled back to his feet and tore off his shirt, swearing and swinging punches. De Lima waved his shovel as Williams swung at him wildly. Punches were exchanged, and rocks were hurled. Eventually the men were prised apart by their companions, and although serious damage was prevented it was clear to all that their working relationship was finished. While Julius Eggers and

Tack Long at Three Mile were old rivals, Williams and de Lima were ostensibly partners, part of a multi-ethnic group of miners working at the Humbug Hill Sluicing Company, which we met in the previous chapter. The two had collaborated since 1856, but were now soon to appear together in the Police Court answering charges and counter-charges of assault. The judge dismissed all the cases, telling the men to sort it out themselves.[4]

At the core of all these disputes was water. On the goldfields of Victoria, people fought over control of water for all sorts of reasons, but greed, ambition, money and theft were typically at the heart of the matter. More water meant more gold.

Eggers and his German mates had been arguing for months with Tack Long and the other Chinese miners. Both parties wanted access to the water coming down a race from Stanley with which they could work their claims along Three Mile Creek below Beechworth.[5] Plenty of water tumbled down the creek for most of the year, and the tangle of races nearby diverted even more, yet there never seemed to be enough to keep everybody happy. It all came to a head in that pitched battle late one afternoon in November 1865. The two groups clashed over water all day, diverting it back and forth between sluicing claims. Tempers rose, shouts turned to blows, and 'the great fighting case' between German and Chinese miners ended up in the Beechworth Police Court a few days later.[6] The judge fined Tack Long £5 for his aggressive role in the affair but otherwise dismissed the whole business. It was a rare legal win for the Chinese.

Over at Creswick, the Humbug Hill workers came to blows over how to use the water they already had. Williams and his three mates wanted to use it in their race to work their claim, as they had done effectively for four years. They were miners and they were happy being miners. Jacinto De Lima and Thomas Lake, however, together with partners John Boadle Bragg and Charles Lewis, saw a new opportunity they were keen to exploit: they wanted to extend their race to carry water to the dry diggings seven miles away at Bald Hills, where they could sell the water to other miners at a profit. The partners had been arguing about the plan for weeks and had already lost out on contracts. One winter morning Bragg and his group were sluicing part of the claim when Williams and his mates seized control of the race and shut off the water supply. De Lima bashed Williams' head against a tree and the fight was on. In the end, Bragg, de Lima, Lake and Lewis bought out Williams and three others, James Videan, Domingo Franciso and John Keating. The Humbug Hill Sluicing Company built the extension and prospered for years as one of the major suppliers of water on the Creswick field.

Bragg and his remaining partners were part of the new entrepreneurial breed making money by controlling the supply of water. These water merchants started to appear across the Victorian goldfields during the mid-1850s, as we saw in the previous chapter, investing heavily in the infrastructure of water supply. They worked out where water originated and where it was needed; they puzzled out the terrain so the water would flow through the channels; and they built dams, tunnels and flumes to expand their networks. But, as challenging as these technical problems were, the legal issues of water control were just as tricky – if not more so. The challenges of water diversion included securing money for pipes and blasting powder and employing labourers to dig the ditches, but no one was going to outlay those monies unless they had secured ownership of the infrastructure they were building and the water they wanted to sell.

The water merchants themselves were part of the problem. Secretary for Mines Robert Brough Smyth pointed out that their initial privileges were only intended as easements, not actual *titles*. Such men soon came to believe, however, that they 'had an exclusive right to the enjoyment of water over which [they] had had control for a long period [and] did not hesitate to sell it'. The claimholder became 'an owner of water'. But water bosses were under no obligation to sell water at a fair price, and Smyth observed that they were jealous of their privileges and looked upon with distrust by other miners. They believed they had 'an equitable if not a legal right to the water' they controlled, leading to endless disputes and litigation.[7]

So, who actually owned the water? There was a lot of uncertainty around this question. Under the British tradition of riparian law that applied in the colonies, the right to use water belonged to those who owned the land beside the river or stream in which the water flowed.[8] Water belonged to a community of users with equal rights and obligations. Their riparian right was one of usufruct, the right to enjoy the use of a resource, rather than ownership, and they could not pollute the water or diminish its flow. In the first two decades of colonisation in Victoria most land was occupied by squatters, and although they controlled access to creeks and rivers, the riparian principle prevented them from diverting or developing water except for stock and domestic purposes.[9] In short, water could be used by riparian owners or occupiers but not possessed in the same way as land or gold.

When John Batman and his party arrived in Port Phillip in 1835 they claimed they had 'bought' the land by exchanging gifts in a ceremony with elders of the Kulin nation. The treaty was quickly rejected by Governor Richard Bourke, who lost no time in asserting that the land belonged to the British Crown, not to Batman or the Kulin.[10] Aboriginal people were officially dispossessed and access to land was managed by the colonial government. The appropriated territory couldn't be purchased from the Crown until it had been surveyed, and most of the country was occupied by pastoralists on simple licences or leases. Only blocks in urban areas and a few

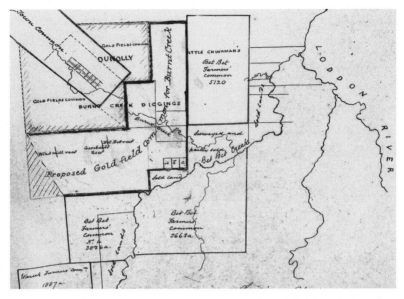

Dunnolly goldfield commons, 1862

special surveys had been purchased outright. In 1854, pastoralists were given the opportunity to purchase part of their leases as 'preemptive rights'. Land sales for smaller farms didn't take place until after the passage of the Land Acts in the 1860s.

During the initial years of the gold rush and for many years thereafter miners worked mostly on Crown land. Once land started getting sold off, the government set aside large parts of the mining regions as 'goldfields commons' that were, initially at least, reserved from sale.[11] Marked out from 1861 onwards, most of these eighty or so goldfields covered a few thousand acres, but a few were much larger, including Inglewood (50,096 acres), Bendigo (83,702 acres) and Castlemaine (22,642 acres).[12] The commons surrounding or adjacent to towns on the goldfields proved to be a valuable public resource for many years, providing firewood, grazing land and recreational space. They were also a haven for Aboriginal people who had survived dispossession and continued to live on country throughout the gold-rush years.[13] Most of the commons were whittled away by land sales over the years before being abolished entirely, but some lingered into the 1940s. Today, the remnants of those areas are the basis for a mosaic of state and national parks that extend across central Victoria, and they remain a valued public asset.

The commons, however, did not resolve the water problem. Mining continued to take place on public land, although only the Crown had official riparian rights to the water that flowed there. The question remained: if the Crown owned the land, who could use the water and cash in on the potential fortune it facilitated?

In the 1850s water was hot property and people went to great lengths to get it. Some, like John Williams and Jacinto de Lima, Julius Eggers and Tack Long, tried to solve the problem with their fists. Others turned to theft, although they didn't necessarily see it that way – they called it 'cutting a race'. Late one autumn night in April 1861, race-keeper George Mason noticed something odd about the flow of water in the race he was tending in the bush near Creswick. The water had been running well, but had slowed quite suddenly to a trickle. Although it was after midnight, Mason followed the race upstream to investigate. Sure enough, someone had shovelled dirt into the race to block it, and then cut a break in the wall of the channel to divert water into the sluicing works of a rival company. When confronted, the thief boldly admitted it, and declared he would do it again if the break was repaired. The owner of the race, John Roycraft, hauled the offenders into court.

Roycraft's rivals were the prominent Russell family. George Russell, born in Dublin c.1815, came to Creswick in the mid-1850s, and his younger brother William had arrived a few years earlier after being transported to New South Wales as a convict in the 1830s for burglary.[14] The Russell brothers quickly developed a range of mining and water interests. It was George's son, George Jnr, who was now being accused of taking Roycraft's water so brazenly. It turned out that when Roycraft had taken over the race from the Eaton brothers a year earlier he had also stepped into an old feud between the Eatons and the Russells. In a typically convoluted arrangement, Benjamin Eaton had stayed on as manager of the race, using some of the water himself, while leasing the rest to Chinese miners who in turn sub-let it to others.[15] It was a tangled web, and, adding to the confusion, several of these race systems were interconnected. Contests over this water had played out in the courts for years and now the younger generation was taking the law into its own hands. When questioned in court George Jnr claimed it was all a misunderstanding. He hadn't been threatening to cut *that* race – he'd meant to divert another one where they had a clear right to the water. Mason of course had not seen who had cut Roycraft's race, so the case was dismissed.[16]

Young George Russell Jnr was hardly the first to try this game around Creswick, nor the first to get off lightly when apprehended. In a complex and litigious era when water rights were up for grabs and no one had a clear idea of who was entitled to what, 'cutting a race' was just one more way of asserting prior access. James Adamson thought nothing of getting up in the middle of a rainy winter night to go out and cut his neighbour's race so the water would flow to his claim. Adamson told the court that as far as he was concerned he was entitled to the water; he'd purchased a right at an earlier date than his neigh-bour. Adamson had apparently assumed that California's doctrine of prior appropriation applied in Victoria. The court disagreed with his interpretation but let him off with a nominal one-shilling fine.[17]

A miner named Bishop was another who tried his luck in the night. One of the earliest races in the Creswick district ran just above his claim. Owned by William Mitchell and James Robertson, it tapped a steady supply of water from the Bullarook Forest to the South-east. It also captured most of the runoff that would have gone into Bishop's dam. When they noticed that the water ran well during the daytime but seemed to go missing at night, Mitchell and Robertson became suspicious. They installed a night watchman, A.F. Hitchcock, who soon discovered what was happening. On his sentry, Hitchcock spied Bishop installing a board across the race, diverting the water into his dam. With this testimony, Mitchell assumed it would be an open-and-shut case, but, when pressed, Hitchcock couldn't be absolutely sure that it was Bishop he had seen in the darkness, so the case fell apart in court. When Bishop produced a witness, who claimed he'd been in his tent all night, the complaint was dismissed.[18]

Conflict over water was frequently exploited as a tool to victimise Chinese miners. Anti-Chinese sentiment was widespread on the goldfields, and the Chinese were often accused of using excess water beyond their licence allowance, encroachment on claims, illegally extending tail races, non-payment of water fees and fouling water with sludge and debris.[19] In 1855, Local Courts at Bendigo and Castlemaine passed by-laws that declared waterholes preserved for domestic use had to be marked by sticks in the shape of a cross so that Chinese miners would know the water was reserved.[20] In 1857 the Chinese were specifically targeted in a regulation that made it illegal for them to destroy, damage, pollute or obstruct any dam, race, sluice or watercourse, or to otherwise interfere with water for mining. Disputes often turned violent, with numerous recorded cases of assaults and punch-ups and the deliberate wrecking of sluice boxes.[21]

All these complaints were, of course, just as frequently made about European miners, as the Creswick cases illustrate. In fact, Chinese miners were far more likely to be the victims of theft and personal violence than the perpetrators. Tack Long and his friends were actually rightfully defending their water when they fought the Germans at Three Mile – an earlier court decision had awarded it to their party. The Germans were unhappy with the ruling and were caught red-handed trying to elude it. Securing the water was a moral as well as a legal victory for the Chinese, even though Tack Long was fined for assault.

Conflicts over water played out against a general backdrop of fear, antagonism and outright racism. When Julius Eggers and his friends took up shovels and forks against Tack Long's party they were using bullying and violence to reclaim what they had lost in the courts. Both groups would have been well aware of the riots some years earlier in the nearby Buckland Valley. In July 1857, 2500 Chinese men had been driven off the Buckland diggings by a murderous mob of white miners.[22] The mob rampaged through the diggings for hours, looting and burning homes, businesses and the Chinese temple. They beat the men and drove them into the freezing Buckland River. At least three Chinese men were killed, but there may have been more among those who fled the valley into the snow and ice of the high country winter. Many of the Buckland Valley dispossessed went to Beechworth, so Tack Long and his friends would have been acquainted with those involved; indeed, they might even have been in the Buckland Valley themselves. The Germans may have been outnumbered in the fight at Three Mile Creek, but their Chinese opponents would have seen the lurking menace behind the Germans' actions.

Violence and theft were not the only ways to secure water; in fact, they were generally disdained. Instead, both Chinese and European miners alike tended to seek justice through the legal system. The generally law-abiding nature of Australian goldminers was noted by English writer Anthony Trollope when he toured the goldfields in the late 1860s. Notwithstanding the punch-up at Three Mile, Trollope observed that miners, 'do not fight and knock each other about. They make constant appeals to the government officer ... and they not infrequently go to law.'[23] The Beechworth Local Court, for example, handled just over 100 cases between October 1856 and March 1857. Of these, at least seventy-nine involved disputes over water. They ranged from failure to deliver agreed-upon volumes to illegally building a dam across a creek, and allowing 'Water to run to Waste' when it was scarce in summer.[24] On remote goldfields, however, miners could be a long way from where the court held its sittings, so disputes had to be settled by mutual agreement – or physical force. On the whole, though, those wanting to use water fought their way through case after case, legal challenge after legal challenge, as the judges, lawyers and legislators gradually worked out where the problems lay. Eventually, after a decade of litigation, they hammered out a system that worked.

The conflict at the heart of the water issue was how to reconcile the commercial use of water with the prevailing understanding of water as *res communes*, a resource collectively owned, like air or light, and available for the benefit of all. The colonial authorities and many miners initially took their cues from English custom and common law, where the approach to determining access to this 'moveable, wandering thing' was based on ownership of the land adjacent to rivers, streams and lakes.[25] Landowners could use the water that flowed past their property for personal use and watering their animals, but they didn't *own* it, and they could not pollute it or diminish its flow. This was the riparian tradition of water rights that recognised shared ownership and a variety of complex and often competing needs.[26]

England, of course, is a 'wet country', to use historian Michael Cathcart's evocative phrase; it is damp, while Australia is dry.[27] Water in England was generally abundant, even as the Industrial Revolution increased water demand to service power, transport and the growing needs of towns and cities. Putting water to work for industry led to increasingly frequent conflicts between users. Gradually, throughout the eighteenth century, a new principle of appropriation emerged based on the notion of 'first' or 'prior' use.[28] The idea underpinning this model was that first usage created ownership, so water came to resemble private rather than common property. The great legal commentator Sir William Blackstone argued that rights to flowing water should follow occupancy, or first possession: 'If a stream be unoccupied, I may erect a mill thereon, and detain the water; yet not so as to injure my neighbour's mill, or his meadow; for he hath by the first occupancy acquired a property in the current.'[29] *First use*, not just use, was the decisive factor. Later users could also partake of water supply, but not if it interfered with those already consuming, and it didn't matter if those first users were upstream or downstream of later users. Time, rather than space or location, organised the hierarchy of usage rights. This notion of prior appropriation was also to become particularly important to the way miners in the American West determined who had access to water.[30]

In the English context, location still had a role to play, and only adjacent landowners had the right to appropriate water. But the question in Victoria was: what do you do in a place where most of the land is held by the Crown, and where miners need to divert water away from the stream bed? In pastoral districts riverside land had been taken up by squatters purchasing homestead blocks under pre-emptive right. On the goldfields few actually owned land or could claim riparian rights. On the Ovens, for example, when John Reilly wanted to build a race to extract and use water from Nine Mile Creek, he went to his local goldfields commissioner, John Morphy. In the early days of the gold rush the commissioners had ultimate authority, and with little to guide them these officials often made decisions on the fly. In this instance, Reilly was effectively drawing on the principle of prior appropriation to request first access to water. He was forcing the issue because he was keen to take advantage of a new mining regulation gazetted in April 1853 that allowed 'Sluice Washing ... at running streams' for the first time.[31] Sluicing relied on water supply, but the regulation made no mention of provision for water storage or diversion, so Reilly decided to try his luck. In the absence of any law to guide him, Commissioner Morphy had to make a call. He granted Reilly a verbal permit, followed up by a written permit issued the year after, in 1854, when such things came into effect:

> Permission is hereby granted to (W.X.) and party, and (Y.Z.) and party, to take two sluice-heads of water night and day, with right to a race, cut conjointly, from the Hurdle Flat Creek and Springs, commencing a short distance below the dairy station, on condition that it does not interfere with the interests of the public.[32]

Early permits like these were nailed to a tree in a prominent location so that they could be seen and read by all.

Reilly may have drawn on prior experience of mining and water regulation in New South Wales after crossing the Pacific from California. In New South Wales, goldfields officials were often hamstrung by government reluctance to clearly define their policy on miners' water rights. Commissioners were instructed to prevent monopoly and regulate supply, as well as to permit new races but only if it was in the public interest to do so.[33] The NSW parliament legalised the construction of races and tunnels in December 1852, but these rules were intended for diverting creeks and draining auriferous land, not sluicing.[34] Regulations applied in February 1853 (earlier than those passed in Victoria) permitted the creation of 'reservoirs for washing gold', but such dams were reserved for the exclusive use of the applicant and were not to be detrimental to the public interest.

Melbourne authorities had no clear idea of what to do about the water problem either. Faced with growing public demand for a decent water supply, the goldfields commissioners responded to local circumstances as best they could. Morphy's colleague at the nearby goldfield of Stanley, Captain Virginius Murray, preferred to hear disputes while tucked up in bed, an eccentric albeit reasonable choice given the bitter cold of winter in the mountains. To make his job easier he drew up an informal regulatory code, and, while its legal status was unclear, it gained a certain measure of acceptance among his fellow commissioners in the region.[35] In the absence of any clear direction from Melbourne they didn't have much choice. The *Act for the Better Management of the Gold Fields* 1855 resolved many of the anomalies and injustices that had spurred the Eureka rebellion at Ballarat late in 1854, but it didn't include any mention of water supply, despite the Gold Fields Commission of Enquiry hearing many complaints on the subject.[36] What the new Act did do, however, was to delegate authority to elected Local Courts, effectively giving miners almost complete control of regulation in each district.[37]

Members of the Beechworth Local Court gathered in a dusty blacksmith's shop in October 1855 to devise a set of regulations that would serve the district for years to come.[38] The rules tended to preserve the field for the small miner, except when conditions dictated large claims. Perhaps most significantly, races were now recognised as private property, with 'A ticket to be given securing to the holder full possession of said race as property'.[39] That was enough for the miners. The work of building water races surged ahead. All over Victoria miners and investors hired gangs of men to dig ditches and build dams. Once these were built, the networks and permits themselves became valuable assets. They were bought and sold for hundreds, or sometimes thousands of pounds, and the mining industry thrived. Part shares in water networks were also frequently traded. Beechworth continued to take the lead in water matters, and within a few years the district boasted water infrastructure valued at nearly £200,000.[40]

In 1857, the Victorian government responded to the situation with its first real attempt at water regulation. The *Act for amending the Laws relative to the Gold Fields* included the right to construct and use races and dams as part of the privileges associated with miners' rights.[41] The miner's right was a simple licence that allowed the holder to dig for gold, vote and occupy a small patch of land for a house and garden. This was the beginning of widespread home ownership in Australia. The provision for water races in the miner's right quickly proved popular, with 200 water rights taken out by miners at Yackandandah alone within a few months.[42] But the Act still relied on Local Courts to establish mining by-laws that met the needs of each area. The complex division of water entitlements, particularly around Beechworth, quickly proved too much for the system to handle: litigation was endemic and the whole shaky edifice fell over less than a year later with a landmark decision by Judge Thomas Cope in August 1858.[43]

Cope was the chairman of general sessions in the Beechworth Court of Mines, but he had only recently arrived in the district after some years in Ballarat, where he had aided the defence of Eureka prisoners.[44] He knew about Victorian mining regulation and, having been educated in Cornwall, was also likely familiar with the British system of stannary law that governed tin and copper mining. But the legal tangle that had arisen from the custom of ad hoc and unofficial legislation in Beechworth was anathema to him, and his decision in the landmark case of *Moysen v Hopper*, while controversial, laid bare the problems with the 1857 Act and regulations.

The case in question was ostensibly a simple dispute over access to water in Alfred Moysen's race at Hurdle Flat. It quickly escalated into a more general debate about the legal ownership of races and the water within them, with big implications for the mining industry. Specifically, section 3 of the 1857 legislation authorised miners 'to divert and use for mining purposes any water which Her Majesty could lawfully divert and use'. However, Justice Cope believed that without legislative authority for doing so, Her Majesty actually had *no right* to divert water from a stream when the adjacent land was privately owned or occupied. Such were the rules of the riparian tradition. Cope was puzzled by the way Beechworth miners had applied the principle of prior appropriation and its consequent ordering of first rights, second rights, third rights (etc.) to water, which in his view was 'a thing unknown to law'.[45] He thought the whole matter in need of referral to the Supreme Court.

Suddenly, all the water permits issued in the colony were in jeopardy. Men who had spent thousands of pounds building water infrastructure now feared their investments may be worthless. The situation was a mess, and an atmosphere of combustible outrage descended almost instantaneously.

Only a few days after Cope's ruling on *Moysen v Hopper*, a large number of water-permit holders met in John Wallace's Star Hotel and Theatre at Beechworth. If their races, the result of years of hard work and investment, were indeed illegal, miners vowed to seek recompense. Race holders kept thousands of workers employed, so terminating licences now could only be an act of the 'grossest injustice'. Donald Fletcher pointed out that working men should not be expected to determine knotty points of law. If their water rights were invalid it was the fault of 'bungling legislators'.[46] The conference debated and passed motions in support of water rights and organised a delegation to Melbourne to lobby ministers. The miners knew Cope's ruling was robbing them of their water rights.

Among government authorities opinion varied as to the impact of Cope's decision. Members of the 1862–1863 Royal Mining Commission had their doubts about the decision but acknowledged it still had the effect of destroying the value of water permits, which were no longer worth the paper they were written on. The privilege of the miner to extract water, they wrote, 'looks almost like a mockery'.[47] Secretary of Mines Robert Brough Smyth was more ambivalent. He suggested that Cope's decision had an impact in places where there was likely to be opposition to water diversions, but elsewhere races were built and rights were established under local by-laws in the old way. Miners clung to the belief that they had at least *some* rights to water and looked forward to a time when the laws would be amended.[48]

Everyone involved with water suffered uncertainty, but it was particularly bad around Beechworth. The issue for miners on the Ovens goldfield was the sheer scale of water diversion and the unusual local practice of sourcing water from groundwater springs in addition to surface runoff from rains. There was so much competition to capture and exploit water from the same sources that the system was grossly inefficient. The old lease plans, which are still archived today in Victoria's Public Record Office, show water systems that look like spaghetti, with many intermingled races crossing over and under each other (see figure on page 121). John Pund's is a good example: in some spots it runs beside up to a dozen other channels, often only a few metres away from them; in its journey westward from Stanley to the Three Mile diggings it crosses other races several dozen times.[49]

We found one of these crossings in the bush with our co-researcher Jodi Turnbull. She used GIS to build up layers of old maps and aerial photography, consulting the composite image to identify where one race crossed another. We parked the car at the point where the race intersected the road, and from there we followed the race into the scrub. When we got to the crossing place we found not only the race but the wooden flume that carried the water in a little aqueduct over the race below. Some of the boards had collapsed but the flume was remarkably intact. Through the grass and blackberries we could

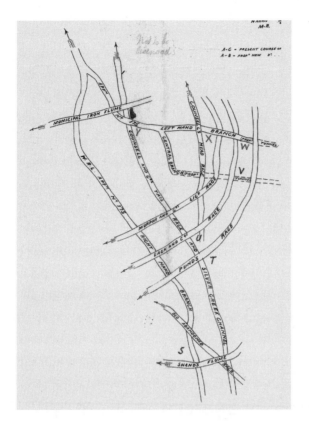

Donald Fletcher's race network

see an open wooden trough about one metre long and half a metre wide, with four narrow boards at the base. The vertical boards that formed the sides of the box had fallen in and the race was eroding. Someone – maybe John Pund's son Percy – had probably repaired the boards over the years, but the route of this antique flume was almost 150 years old.

Duplication of water races was a waste of time and labour, and it multiplied water loss through seepage and evaporation. Mining entrepreneur and water merchant John Alston Wallace sensibly pointed out that it would be better and cheaper for everybody if water was delivered in one channel rather than six or eight or more. This would have required, however, a degree of cooperation and trust between mining parties that was usually lacking.[50] Water claims often exceeded available supply, verbal entitlements were confused or forgotten and paper permits were lost. It became increasingly difficult to prove who had rights to what in this shambolic system.

The Ovens goldfield also featured a mix of different *kinds* of rights that created further conflict. 'Creek rights' took first precedence. This meant that a single miner could put a sluice box or cradle in a creek and then lawfully demand that water continue to flow in the stream. By-laws specified that at least one sluice-head of water had to be left running in streams to supply these 'tub-and-cradle men'. Miners with 'bank rights' – that is, those who had spent large sums and many months to dig races – had to comply with creek-rights rules or risk a summons to the Police Court and hefty fines. It was even more galling when just one or two miners held the creek right, and thus had first call on the water.[51] Those with 'motive power rights', who wanted to use the water to run machinery such as water wheels, were at the back of the queue and had to get by with what others left behind.

Beechworth miners also used tail – that is, waste – water released by their upstream neighbours. Small dams were dug to take advantage of the muddy water flow, and this source was often used by several successive claims further downslope. Waste water was regularly stored overnight and used the next day. Water prices reflected this. A sluice-head could jump from the standard £2–3 to £10–12 if the original seller knew that the purchaser would be able to onsell the tail water to a succession of parties downstream. James Pendergast used the water in his own race below Stanley and then let it to others for £13/10s per week, before it flowed away to Yackandandah.[52] Water rights were sometimes divided into day and night rights to allow for continual use. By 1858 on the Ovens goldfield this was regulated to the point that miners had to apply to the warden for permission to use others' tail water as if it were creek water.

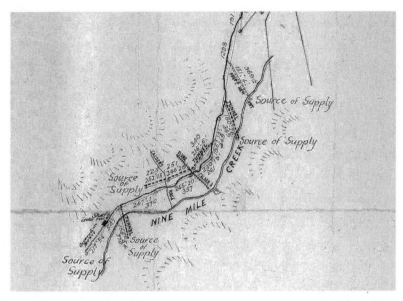

John Pund's sources of supply

Underground or spring water was also a unique issue around Beechworth (see figure above). Engineer Joseph Higgins identified sixteen well-defined springs between Beechworth and Stanley for the Ovens Gold Fields Water Company in 1860, water that was otherwise 'running to waste'.[53] Miners began excavating horizontal tunnels deep into hillsides to tap groundwater springs and increase the supply of water flowing into their races. Many of these 'source of supply' tunnels were hundreds of metres long, often with four or five tunnels supplying each race.[54] The digging was hard. William Walker, the race-keeper on the Big Ditch at Yackandandah, spent months blasting and cutting slabs to drive a tunnel.[55] To avoid disputes, each tunnel had to be at least 100 yards from those belonging to other parties.[56] The amount of water flowing from each spring varied, but they provided a useful supplement in summer months when creeks ran low or dried up altogether. Some Beechworth water users claimed that by digging tunnels to tap underground sources they were 'creating new water' and therefore deserved stronger protection of their access privileges.[57] Others argued that such water was merely part of the natural creek supply.[58] Inevitably some miners dug tunnels to tap springs at a higher level and deprived those below of the water they had secured. It all led to more conflict.

Water contests between miners and local councils also spilled over in the form of civic and legal disputes. In 1862, the Beechworth municipality completed a town reservoir named Lake Kerferd after local brewer and councillor George Briscoe Kerferd. Two mining groups, Connolly & Co and Edward Spencer and Samuel Gaylard, had licences affected by the reservoir, so the council agreed to compensate them by providing each party with three sluice-heads of water. Alarmingly, it soon became clear that even in winter the reservoir only held enough water to satisfy the miners' entitlement, with little left for Beechworth residents. The reservoir's capacity was actually shrinking as it filled with sludge from workings in the eastern part of the catchment. Reports and surveys over the next ten years failed to either solve the dispute with the miners or increase supply, and, finally, in 1873 the Beechworth council simply shut off access to the miners and kept the water for town residents. Connolly & Co sued. The case ended up in the Supreme Court, where Justice Robert Molesworth found in the company's favour and ordered the Beechworth council to pay costs.[59] Connolly & Co accepted £4250 in compensation; Spencer and Gaylard settled for £550.[60] With the council still deeply in debt for bungled waterworks projects, this was an expensive way for ratepayers to test the limits of water supply.

In nineteenth-century Victoria most mining took place on Crown land but there were many occasions when auriferous country clearly extended into privately owned land. Mutually beneficial arrangements were made in these cases, with governments turning a blind eye and wardens refusing to deal with disputes that occurred around mining that wasn't on Crown land.[61] One of the best known and most successful examples of mining on private property was the Port Phillip Company. The company negotiated to mine on privately owned land at Clunes, but even here the legal status of the agreement was dubious. When other miners tried to tunnel into the property from outside its boundaries the company had to defend itself on the ground. There were pitched battles and hand-to-hand fighting in the underground tunnels.[62] The solution was obvious to everyone: legalise and regulate mining on private land and compensate landholders for damages, by agreement or adjudication. Parliament tried to resolve the problem numerous times, but it was not until 1884 when the *Mining on Private Property Act* was passed that the issue was ultimately settled.

Prior to that, things could get tricky when a miner wanted to cut a water race or tail race through private land. When the Crown wanted to enter private land to secure some public benefit – building a road, railway or pipeline, for example – there were regulations about easements and rights of way to facilitate this. Likewise, a miner could cut a race through unoccupied Crown land without hindrance. If the land was subsequently sold, the miner retained the right to enter the land for the purposes of keeping his race 'in efficient repair', and to dump any accumulated silt to the side.[63] Building a race that ran through private land, however, required agreement from the landholder, who could insist on compensation or just say no and refuse access. This was part of a wider problem of mining on private property that took decades to resolve. Under the doctrine of royal minerals, all gold (and silver) belonged to the Crown, which made gold mining illegal without a licence that could only be issued by government. Any agreement between miners and private landholders to dig for gold was technically unlawful. This posed real obstacles in New South Wales, where much of the goldfields had been sold before mining began, and it's one of the major reasons for differences in the way the two colonies administered their goldfields.

After Justice Cope's decision was handed down, miners demanded action to more emphatically clarify the rules surrounding water. They needed to provide greater security for investors, and for themselves. The government responded with a series of inquiries.

The first was a Commission of Enquiry in Beechworth itself in 1861, where members reviewed water rights and their regulation in the district.[64] The commissioners heard abundant evidence about the inconsistency in water-permit conditions, concerns about the legal status of permits in the aftermath of Cope, and confusion about groundwater rights. Much of the trouble stemmed from the English habit of applying custom and precedent to water governance, which was proving totally inadequate in the new circumstances of a colonial resource boom. It was time to replace ad hoc decisions with a clear, uniform and codified set of laws that were transparent and consistent.

The result was the 1862 *Act to amend the Law relating to Leases of Auriferous Lands and for other purposes*, a formal enshrinement of the government's power to regulate water for mining purposes. One of the Act's first changes was confirmation that races, dams and reservoirs were articles of property that could be bought, sold and leased. The Act also clarified control of the water itself: miners could now 'take or divert water from any spring lake pool or stream' on Crown land, which meant groundwater could be captured as well as surface water.[65] Water licences were issued for up to fifteen years and could be permitted for tail races and reservoirs as well. These races had to be accurately surveyed and plans lodged, with easements of up to ten feet (three metres) either side of a race. The land used for a race had to include a 'gathering ground' of up to four acres for every mile in length of the race, which equalled a land width of thirty-three feet (ten metres), although this was a flexible measurement to allow for curves in their course.[66] Annual rents of at least £5 were now payable. Unlike today's water market, rent was for the land the race occupied – there was no fee for the water itself. In other words, water merchants made their fortunes by selling a natural resource they obtained for free. It was their work in transporting the water that gave it value, and their costs in providing the infrastructure that they passed on to customers.

To apply for a water-right licence, miners had to mark out the course of their proposed race with posts three feet high and set no more than 100 yards apart.[67] The top of each post had to be painted white. They then had to deposit £10 with the clerk of the mining warden as a fund for the costs of survey and submit a written application to the minister of mines. A mining surveyor then went out to appraise the proposed route and prepare a detailed plan, noting sources of water supply, older races, road and track crossings, nearby property boundaries, flumes over gullies and any springs that increased the water flow. These plans were drawn up on large map sheets at a scale of eight chains (1:6336) or even four chains (1:3168) to the inch, with summary information about the length of the race, the area of land reserved as the gathering ground and the total daily volume of water to be diverted. Three weeks were allowed for objections, with the application advertised in the *Victoria Government Gazette* and a local newspaper before the licence was granted. Water-right licences could also be sold to another party. All this administrative process was a far cry from the verbal agreements and scraps of paper handed out as water permits by goldfields commissioners ten or twelve years before.

The Victorian Public Records Office retains some of these remarkable water-right plans from the Beechworth district, mostly dating from the early 1880s.[68] Like any well-drawn map they include a wealth of information about the shape of the land and how people sought to exploit it. The plans detailed the plotting of distances and bearings taken in the field by the local mining surveyor, following the approximate course of the race initially marked out by the miner applying for the licence. The result was a long wriggling line that shows the race following a contour along the hillside. Survey backsights and foresights were marked carefully in blue and red ink respectively. Houses, schools, races and Chinese market gardens were also indicated. Each plan included a standard statement signed by the mining surveyor endorsing the accuracy and quality of his work. 'I hereby certify that I personally surveyed the race represented by this plan', read the inscription by Henry Davidson, mining surveyor at Beechworth, on each of his plans, and

> that the instruments used were a theodolite and chain both in perfect adjustment; that the bearings are taken from the magnetic meridian, the variation of which is given on the plan; that the survey is connected with the nearest available fixed mark in the locality; and that the District Surveyor was furnished with a certified copy of the plan immediately after its completion.[69]

Surveyors like Davidson, with their authority to determine the physical boundaries of virtually all mining activity, wielded unusual power. In the early years of the gold rush, miners had taken it upon themselves to measure up the extent of their claims, leading to fights on an almost hourly basis.[70] Ballarat miners saw the wisdom of employing professional surveyors around 1855, and the practice soon spread to other goldfields, with individual surveyors later approved by district Mining Boards. By 1857 there were eleven such surveyors in Ballarat. They had many masters, including the Mines Department, the Court of Mines, the Geological Department and the district Mining Board, but in effect they were accountable to no one. One of the objections to their role was that their formal duties were undefined. Another problem was that, despite being government employees, they were paid by fees rather than salary. The standard fee for surveying a race, for example, was £3 per mile, with travelling expenses of 4 shillings per mile. Survey of ground for tunnels was also £3 per mile, with the provision of a plan, tracing and report included as part of the fee.[71] A busy and energetic mining surveyor could prosper. But those in quieter or fading places earned much less than those on crowded, productive goldfields. Of course, there was also concern that the surveyors' independence and fee-for-service led them to partiality, and even to extortion and fraud.[72] On balance, mining surveyors and the maps and plans they created probably resolved and prevented far more fights on the goldfields than they ever started.

Mining surveyors often also served, simultaneously, as mining registrars. The position was administrative and involved keeping the books on registrations, shares, certificates, water rights and so on. Individuals like Peter Wright at Yackandandah, James Stevenson at Creswick, Ferdinand Krausé at Ararat and Mark Amos at Fryers Creek were executing both roles by the 1870s.[73] In effect, this meant that mining surveyors went out in the morning to survey a claim and returned in the afternoon to log the paperwork. This was particularly useful when, as so often happened, the surveyor was summoned to court as a witness to swear on documentation he had prepared with one hand and registered in the office with the other. Many were also appointed as inspectors of mines under the *Regulation of Mines Statute* 1873, which established safety rules for ventilation and blasting.[74] There were probably no other figures who knew as much about what was happening on the goldfields.

Not only are the old water-right plans revealing historical documents but we can use them to help identify the remains of races and other water infrastructure preserved in the bush today. High-resolution scans can be georeferenced and aligned to modern maps with GIS software. Control points such as property boundaries or roads are added that link the plans with an underlying digital base map. The original maps are then 'stretched' electronically to create a best fit, and so old and new map are effectively sewn together.[75] Races and dams can be traced and superimposed on topographic maps or Google Earth to see where they lie in relation to modern features. The georeferenced plans can also be imported to a tablet or iPad and used in the field as an alternative to GPS. In other words, we can literally walk around a goldfield, following an old map, with a little blue dot on the screen showing exactly where we are. We've used this method to walk parts of John Pund's race and to identify many of the races around Creswick and other goldfields.

By the early 1860s the government had finally acknowledged the significance of water security for the success of the mining industry, and the regulation of water was now firmly embedded in colonial law and policy. Provisions for mining water were even included in other legislation relating to the regulation of land, including Gavan Duffy's famous *Land Act* 1862. The *Land Act* was the most important contemporary land legislation enacted in Victoria, intended to open up ten million acres for selection. Although rorted by squatters, the Act and subsequent laws began the breakup of large squatting leases, making land increasingly available in small blocks to farmers. The Land Acts of 1862, 1865 and 1869 were the product of years of agitation and the centrepiece of land reform in the colony.[76] Significantly, the original Act recognised pre-existing mining races and reservoirs on Crown land and protected them if the land was sold. Water regulation had finally arrived.

Of course, it wasn't quite that simple. Before the new regulations had even come into effect, a new royal commission of Enquiry was launched in August 1862 to investigate the conditions and prospects of the Victorian goldfields. The commissioners examined more than eighty witnesses and produced a fat report almost 500 pages long, most of it verbatim testimony. Much of the commission's work was taken up with problems of administration on the goldfields, but the twenty witnesses interviewed at Beechworth commented freely on water for sluicing and the inadequacy of existing regulations. Legislators would need to continue working on improvements.

The new legislation they passed in 1865 established the framework that would govern water supply for mining for the next 100 years. The *Mining Statute* of 1865 was a legal landmark, with 246 sections and thirty administrative schedules. It was the distillation of fourteen years' hard-won experience on the management of mining and the administration of justice on the goldfields. It clarified mining laws and spelled out in detail, among other things, water rights and the privileges and penalties associated with miners' rights and mining companies. It restated the previous terms of the water-right licence and established penalties, including fines and prison terms for wrongfully taking water from a race or reservoir. Water rights could be extended from original claims to deliver water, additionally, to any other claim. Pollution of races and reservoirs with sludge or other 'noxious matter' attracted penalties. Holders of mining leases were also entitled to construct 'races drains dams reservoirs' in the same manner as holders of miners' rights.

Miners were more than willing to give these new regulations a try. Almost 100 water-right licences were taken out in the three years or so following the 1862 *Gold Fields Amending Act* – half in the Beechworth district, with another dozen or so each in the Ballarat and Bendigo areas. Most paid an annual rent of about £5 for a race and sometimes a reservoir. As for water-right licences, these were issued in a single running sequence for the whole colony. The first licence was issued in May 1863 to John Fairbairn and George Hunt of the Lal Lal Waterworks Association, and the second to John Roycraft and Benjamin Eaton at Creswick.[77] While most licences were taken out for gold mining, sluicing for tin deposits began in the 1870s in south Gippsland, where John Amey secured a water licence at Agnes River in 1879. The last known water-right licence under the original legislation was issued in 1973.[78] Water-use licences are still permitted under the *Water Act* 1989, however, with the fifteen-year limit still in place.

Numerous applications for water-right licences at Beechworth were not taken up, however, because miners considered the rent too high. In one case an applicant refused to accept his own licence to divert 162 million litres per day because he thought £50 per annum was too much, even though this translated to only a penny for every 30,000 litres of water.[79] The balance between what miners believed they should be charged and the gold they suspected was available was a delicate one, and the government had to seek equilibrium in stimulating the industry while maximising revenue through licensing a natural asset.

The passage of the 1862 *Gold Fields Amending Act* and the 1865 *Mining Statute* were major breakthroughs. These new pieces of legislation allowed individuals rights of access to use water, but they also confirmed that ownership of this water belonged in public hands, for the public good. Significantly, they placed a time limit of fifteen years on access rights rather than an indefinite or perpetual entitlement, as was the case in California.[80] Water was now firmly under government control. After much struggle, Victorian miners had arrived at a unique and effective compromise between riparian and appropriation principles. Based on local conditions and experience, they developed a system that was fair and just for users but also one that recognised the collective importance of water for the whole community of users. It was not without fault, but it was a system that would serve Victorians well. It would also have ramifications far beyond the mining industry.

Water was now also comprehensively commodified; it had become part of the capitalist economy. A sluice-head of water, delivering about a million litres per day, generally cost £2–3 per week. Like minerals, soils, trees and grass, water was part of the suite of natural resources that were exploited by colonists for private gain. It had an economic value that was widely recognised and sanctioned by the community. The infrastructure of dams and races was tradeable as property, while shares in water privileges could be leased and sold. Water-right licences were fully transferrable.[81] There was now a fully functioning, extensive and lucrative marketplace for water.

This new legal framework set Victoria subtly – but significantly – apart from that other great mid-century mining boom across the Pacific in the American West. On the Californian goldfields, miners had faced similar challenges securing water. Like Victoria, their legal system emerged from the English tradition, and their water law, 'the California Doctrine', included both riparian and appropriation principles. When it came to the miners' need for extravagant quantities of water, however, courts promoted use and appropriation over the public good. The basic principle of prior appropriation was *qui prior est in tempore, potior est in jure* – 'he who is first in time is the first in right'. This meant that the first person who came to a stream and claimed its water had priority to exploit it, making the water a form of personal property.[82] This included the right to buy, sell and lease water, and protection for investments in water development.[83] Some elements of the California doctrine were applied on the Victorian goldfields, including the use of water for mining by non-landholders, water trading, seniority of water rights and the forfeiture of unused water.

Where the American miners, in the absence of government authority, enshrined individual rights to property and the productive use of water, miners in Victoria looked to the full array of British law. They used legislation, parliamentary inquiries and court judgements to regulate mineral and water rights and resolve the many disputes that arose. The result was an endorsement of collective ownership and individual rights of access within circumscribed limits. This had significant implications for the way that later water-use regimes developed in Victoria, and, after Federation, in Australia. It provided the foundation for new laws where control of water was centralised and managed for the wider public good.

By 1865 the major battles over water for mining had been won. New mining laws secured title to water through a system of fifteen-year licences. In 1868, Robert Brough Smyth recorded hundreds of privately funded water storages and almost 4000 kilometres of water races that had been built across the goldfields.[84] That's the equivalent of digging a ditch from Melbourne to Perth with a pick and shovel. But the technical marvel of that achievement is overshadowed by the significant cultural shift that the miners achieved. Water merchants changed water from a 'natural good' into a commodity that could be bought and sold. They succeeded in upending centuries of tradition that tied water rights to land ownership. They forged a new direction that was uniquely Australian, unlike traditions in the Old World, and different again from what was unfolding at the same time across the Pacific.

From a British tradition that tied water rights to land ownership, Victorians began to develop a new system that made water a public resource, owned and managed by the government and allocated to individuals under licence. Victoria was among the first jurisdictions in the world to form this distinction. The nationalisation of water and the creation of water rights that emerged on the Victorian goldfields had significant implications for the subsequent development of irrigation agriculture and continues to shape the modern water market.[85] The achievement of those water merchants has a lasting legacy in Australia's management of water – one that can scarcely have been imagined by a man like John Reilly when he requested permission to build his race in the hills above Beechworth in 1853. Fifteen years after Reilly built his race, fights about water on the goldfields were over. But fights about sludge were only just beginning.

5

THE SLUDGE QUESTION

THE SMALL TOWN OF TARRAWINGEE SITS ON THE WIDE flat grazing country between Wangaratta and the foothills of the Great Dividing Range to the east, just off the main road to Bright and the ski hills at Falls Creek. Hodgson Creek flows through the district on its way to join the Ovens River. Before it gets to the Tarrawingee flats, the creek runs through the old gold diggings at Three Mile in the hills behind Beechworth. The archives reveal that, during the nineteenth century, farmers around Tarrawingee complained long and loud about the sludge that Hodgson Creek brought down to smother their land. Without question there had been sludge at Tarrawingee – 10,000 acres of it, according to newspaper reports. Was it still there though? What happened to it in the hundred years and more since the farmers complained? Would we be able to find it?

Our GIS mapping suggested we might, so we decided to try. We used Google Earth to trace the modern course of Hodgson Creek through the floodplain and identify places where the waterway could be reached by road. Then we set off, driving down dusty, ruler-straight roads lined with huge old river red gums. Fat black cattle grazed in lush green paddocks that stretched into the distance on either side. Sure enough, when we reached the end of the road and got out of the car, there was the creek, still carrying a flow of spring rain. Like most Australian streams, Hodgson Creek had once sat level with, or even slightly above, the adjacent land – early explorers called them 'chains of ponds'. In the 1880s people recalled the creek as it had been thirty years earlier: 'large ponds of water very deep, beautifully clear, and with a good supply of fish. The banks were well-grassed and of good fattening quality'.[1] Today, erosion has taken its toll and the stream is now a far wider and deeper scar in the floodplain. The low natural levees that once held it above the floodplain have been replaced by steep cuts in the earth, and the creek runs at the bottom through a deep, wide chasm. Farmers have planted trees and grasses along parts of the channel's edge to hold the soil in place, and catchment managers have built low weirs of rock and concrete to slow the current. Along those reaches the water flows gently through a verdant basin; in other parts cattle wander along the creek bed, erosion is still active and the bare soil is bleakly exposed.

We found a place where recent landslips had revealed the full profile of the bank and there we clambered down to the water's edge. What we saw from the waterside was unmistakable: a wide band of orange soil more than a metre thick, running along the cutting above our heads – yes, there was *definitely* still sludge here. The contrast between the dark organic soil of the old floodplain and the bright orange sludge above it could not have been starker (see picture section). Sediment dislodged by John Pund and his fellow miners upstream near Beechworth was interred here in the riverbank at Tarrawingee. The thickness of the orange sludge layer made it patently clear to us why farmers had been so upset. It must have been impossible to farm – or do much of anything on one's property – with that deep, sticky mud enveloping the paddocks.

Tarrawingee farmers complained to everyone they could about the sludge on their land. It took fifty years of such protests, but in the end these farmers succeeded in doing what few in the 1850s would have thought possible. With other farmers around the colony, their complaints resulted in laws that made miners responsible for the waste they created, and they stopped the creep of sludge over Victoria. By the twentieth century, the new regulations required John Pund's son Percy to build settling dams on Three Mile Creek to keep the sludge from reaching Tarrawingee. Tailings dams like Percy Pund's soon became standard practice in the mining industry. Victorian leadership in protecting the environment from mining waste eventually influenced lawmakers all over Australia and in other parts of the world. The answer to Victoria's sludge question turned out to be some of the first environmental protection legislation in global history.

It was the sheer deluge of complaints from farmers at Tarrawingee and other downstream communities that first drew our attention to the sludge phenomenon and the organised battle to stop it. One of the most vocal of these farmers was Hopton Nolan, a well-known figure in the Tarrawingee district who we've encountered in earlier chapters. Nolan was born in Dublin in 1828, and came to the Victorian goldfields in 1852 alongside fellow passenger John Alston Wallace. The pair briefly went into the hotel business together, and then Nolan ran the Sun Hotel at Nine Mile before he moved down the hill to Tarrawingee in 1860. There he became a farmer, grazier and storekeeper as well as a publican at the Plough Inn, and his daughters ran the post office. By the time the Board Appointed to Inquire into the Sludge Question (hereafter, 'the Sludge Board') came to town in 1886, he was president of the North Ovens Shire.[2] Nolan – 'Hoppie' to his friends – had worked as an alluvial miner on Three Mile Creek in the 1850s, so he knew all about sludge and those responsible for producing it. By the 1880s, though, he had traded in the mountain goldfields around Beechworth for almost 500 acres of good farmland on the rich alluvial flats of the Ovens River, on the floodplain along the road to Wangaratta.

Nolan was scathing about 'the sludge and filth that came from the diggings', claiming it not only inundated and ruined good farmland but destroyed whole belts of trees as well.[3] Over the years the problem had become so serious he was forced to build a two-foot-high levee around most of his property to keep the sludge at bay. When the Sludge Board arrived to see the problem for themselves, Nolan joined a group of men from the district, including County Court judge James Casey and local politicians Thomas Bent and James Brown Patterson, to showcase the scale of the problem.

Tarrawingee farmers were not the only ones complaining. Old newspapers and government files contain endless discussion of 'the sludge question', a phrase that recurred over and over from the 1850s through to the 1920s. Trove, the National Library of Australia's online repository, has literally thousands of stories about the issue from Victorian newspapers throughout the period. Everyone was essentially asking the same question: 'what are we going to do about these endless tides of mud that are burying our homes, businesses, farms and roads?'

The answer wasn't so simple. Few existing regulations could be applied explicitly and effectively to sludge. Like so many other aspects of the gold rush, mining waste was an entirely new problem. Members of the Sludge Board summed it up neatly in 1886 when they said that 'the circumstances arising out of the discovery of gold in the colony were entirely novel to a British community, there having been nothing analogous in the mother country to furnish a guide or warning'. The Board went on to explain how, amid this confusion, the actions of the miners established certain community expectations: 'it came about that from the earliest times the creeks and rivers were looked upon as the natural outlets for the débris from the mines, no thought being given to the ultimate effects of filling up the watercourses. In fact, riparian rights, to a certain extent, have since been claimed by the miners to have accrued through long usage.'[4] What began as an expedient solution to an immediate problem – one generated by industry and with industrial prerogatives in mind – had congealed into an accepted condition of life in the colony.

Piecemeal attempts to regulate waste through successive Mines Acts and in regulations passed by district Mining Boards were attempted from time to time, but these were limited in scope and generally ineffective. Local debates touched on whether laws and customs from elsewhere might be relevant – some suggesting that English drainage acts could provide a solution, or that there may be remedy in the age-old custom of 'natural justice'. Landowners in England had been clashing with tin miners over this same issue in the 1850s – was it a reasonable practice to discharge mining waste into a stream and pollute the water for downstream users? In 1857, the Court of Exchequer had found in favour of industry, judging that, on balance, the custom of mining was so broadly beneficial and necessary to all that the damage to landowners downstream was tolerable.[5] If lawmakers in Victoria were aware of such judgements they held little comfort for property owners inundated with sludge. For decades the issue remained contentious and largely unresolved.

In legal terms, sludge fell under the heading of 'nuisance'. Notionally at least, it was incumbent on miners and others using a stream not to use it in such a way that injured other parties.[6] The Mining Boards of Ballarat and Maryborough, for example, prohibited sludge from flowing onto public roads or other mining claims but were silent on sludge damage to private property.[7] In the Beechworth and Bendigo districts, miners had to build a channel to convey their sludge to the main drain of a gully or flat, and to otherwise keep their tailings 'clear of any other claim, race, or tail-race'. The same rules applied at Castlemaine, while at Ararat miners were required to build an embankment to retain the sludge.[8] Despite these rules, miners and mining companies all over the colony kept discharging their debris into the nearest creek, and although such 'mischief' was grounds for legal action, successful prosecutions seem to have been few and far between. Miners often argued that since their own sludge was just a fraction of all the waste coming down a creek, who was to say which part was theirs?

The struggle to control sludge was a prolonged duel between mining interests on one hand and those fielding the damage on the other. While mining remained the biggest economic game in town, the odds remained in favour of miners. The value of gold outranked wool as an export in Victoria until 1874, and even in the 1880s the industry employed two-thirds as many people as manufacturing. Collectively, far more people lived in the major goldfields towns and surrounding regions than in Melbourne.[9] Many politicians had started out on the goldfields themselves, so they had first hand experience of mining. The industry had the money, the jobs and the ear of government. There could be no answer to the sludge question that got in the way of industry.

And yet, as the century wore on, the odds shortened as other sectors of the economy developed and interests competing directly with mining became more powerful. By the 1880s the value of agriculture had become nearly three times as much as mining. Even in goldfields towns, economies had begun diversifying, with manufacturing and agriculture playing larger roles. Foundries built to supply the mines, like the Union and Phoenix foundries in Ballarat and Thompsons Foundry in Castlemaine, were now also producing machinery for agriculture and the railways.[10] Aided by tariff protection, local manufacture began to replace imports.[11] And, from the 1880s onwards, the government took a serious interest in promoting irrigation agriculture and began to invest in infrastructure for water supply. Another factor was the gradual exhaustion of gold reserves. The next big rushes in

Australia occurred in northern Queensland and Western Australia in the 1890s, far away and at opposite ends of the country.

The long history of protests and responses – not only in Victoria but across the continent in places where mining took place – illustrates a paradigm shift in the calculation of costs and benefits, and in what philosophers call 'enlightened self-interest'. Gradually, the effects of sludge were deemed too costly to the broader economy. This history also shows the slow but steady emergence of an environmental consciousness and a widely expressed lament for the loss of beautiful places degraded by mining.

Initial attempts to control sludge started as early as the 1850s, when rough digging camps started transforming into towns. Business owners were dismayed when their premises filled with sludge, and councils disliked having to clear roads and rebuild submerged bridges. In Castlemaine, the goldfields commissioner John E.N. Bull took the problem so seriously that in January 1855 he issued a proclamation banning puddling machines from the district.[12] This didn't go down at all well with the local mining community, particularly coming only a month after the Eureka protests in Ballarat, where more than two dozen men and women had been killed by government forces.[13] The response to Bull's directive was a large protest meeting followed by a petition to the governor signed by several hundred local miners and business owners. The petitioners highlighted the economic and social value of puddling to the local community; and, in the Eureka aftermath, the government decided to refer the matter to the Gold Fields Commission of Enquiry rather than collecting evidence themselves.

The Commission agreed with miners that puddling should continue: the revenue it generated was far too important. It also decided that governance of the goldfields should be reorganised. In 1855, Mr Bull and his fellow goldfields commissioners were replaced by

Local Courts, with sitting members elected by holders of miner's rights.[14] In turn, the Local Courts were replaced in 1858 by Mining Boards – one in each of the colony's six (and later, seven) mining districts (see picture section). The Boards each comprised ten elected and paid members representing divisions within each mining district, and they were tasked with setting and implementing appropriate regulations for each district.[15] The early goldfields commissioners, government appointees who tended to be members of the colonial elite, had been quasi-military in organisation, wearing uniforms and backed up by the police. The new Boards were far more democratic, comprised local men and were elected via secret ballot by miners themselves. But, though the Boards were more representative and their members understood local conditions, the regulation of sludge was now in the hands of those creating it. Such industry self-regulation produced unsurprising – and perhaps inevitable – results.

The struggle to control sludge in Ballarat was typical of what unfolded in many goldfield towns. The new *Mines Act* 1855 had little effect, and sludge continued to flow. In 1857, the *Mines Amending Act* included an obscure provision in section 111 that empowered Mining Boards to collect fees from miners for keeping sludge and water channels clear. Later that year, the Ballarat Roads Board started taking puddlers to court.[16] Like the Mining Boards, the Roads Boards were new institutions created in the 1850s, with responsibility for building and maintaining roads. Among their responsibilities was the removal of sludge and the rebuilding of submerged bridges, but they were clearly tiring of it. The Roads Board constable of Ballarat issued warnings to puddlers, but they were ignored. Finally, in its frustration, the Roads Board took the offending puddlers to court, where they were issued with fines.

That goaded the Mining Board into action. Amid the winter rains and mud of June 1858, the Ballarat Mining Board decided that mining surveyors should be invested with the power to stop miners releasing sludge that 'interfered with others'.[17] Defined very narrowly, interfering with others only meant damage to *public* roads, *public* watercourses, or *someone else's* mine workings. The Board didn't think damage to other kinds of property was their problem, or miners' responsibility. It soon became clear that even these limited measures were considered to be going too far: new by-laws in 1861 removed the provision about damage to watercourses. It seemed that, as far as the mining industry was concerned, the only real issue with sludge was protecting other miners and, if absolutely necessary, roads.

Industrial self-regulation was not proving very effective. Forced to do what they could to manage the flow, the two municipal councils of Ballarat and Ballarat East decided to shift the sludge somewhere else, and so in 1861 employed engineers to build channels to carry the problem away.[18] The first of these channels were rough timber-lined or earthen ditches intended to confine the sludge in one place and help it more easily flow downstream. Keeping the channels clear and functioning was a constant battle, but they did help to remove the sludge. The battle was won – at least in central Ballarat. Today more than twenty kilometres of sludge channel still carry Yarrowee River and its tributaries through town. Over the years these water passages have been widened and deepened many times and the original timber lining gradually replaced with masonry (see picture section). Now, the massive bluestone and brick culverts and bridges with their elaborate cast iron and steel lattice railings are a heritage-protected, valued part of Ballarat's historic landscape. The creek has become a stormwater drain and its close historical association with sludge has been forgotten.

Over the ranges, Creswick has smaller but similar drains along Creswick Creek and its tributaries. The town's central business district runs along Albert Street, down the hill, and at the bottom, on the flats where the Midland Highway turns off to Daylesford and Castlemaine, the main road parallels Creswick Creek for several kilometres. Today, this is where flooding occurs. In big storms like those of the spring and summer of 2010–2011 the creek floods the entire area, causing millions of dollars of damages to homes and businesses. Though the flats were the heart of the local diggings back in the 1850s, Creswick Creek was not the problem. The real issue was a tributary running down the hill behind Albert Street – an apparently innocuous little creek that carried waste from major diggings further up the hill at Nuggetty Gully. The creek crossed Albert Street halfway down the hill and joined Creswick Creek in the middle of the flats, where the roundabout with the Midland Highway is today. Before it got there it dumped all its sludge in the heart of the Albert Street shopping district and everyone was furious.

Albert Street shopkeepers petitioned to local council: 'owing to the accumulation of sludge [the street] is impassable'. To avoid the sludge, passers-by were crossing to the far side of the street and it was bad for business.[19] All through the winter of 1859 the council debated the sludge question. In the end they obtained a grant from the Public Works Office in Melbourne and used the summer months to start building a sludge channel. The new channel was completed just in time for the next onslaught of winter rains – and hence sludge. The creek was diverted so that it avoided the centre of town, and then rejoined Creswick Creek further downstream.

Alas, while the shoppers could cross the street in comfort, the channel created new problems down on the flats. Now the sludge went directly through the Chinese camp at Black Lead, becoming a nuisance to Chinese shopkeepers and residents. Community leaders, including several European women married to Chinese men, organised their own petitions, demanding several times during the late 1860s and early 1870s that council clear the channel and improve the footbridges. They pointed out that they too 'promptly and cheerfully pay such rates as are from time to time imposed' and should be 'on a footing with our European brother burgesses'.[20] More work was done over the following years and Creswick's sludge channel is now as impressive as anywhere else's. Its drain and culverts, all solidly built in bluestone, are evidence in masonry of just how serious a problem sludge once was.

At Castlemaine, sludge from Chewton flowed west down Forest Creek to flood the town's growing business district.[21] Repeated inundations throughout the 1850s induced the local council to act, and the channelling of Forest Creek began in 1860. A considerable length of creek in the centre of town was moved south, away from the business district, and the meanders were smoothed and straightened. Sludge from Chewton now bypassed the town's market area and flowed into Barkers and Campbells Creeks. Embankments were built along the new channel and it was lined with sandstone to keep the sludge flowing freely.

In 1859 a group of landowners downstream from Bendigo's mines presented the government with a bill for £7600 in property damage and lost income.[22] Calling themselves the Epsom Sludge Committee, the group claimed compensation for 'most ruinous loss' of garden crops, damage to water supply and water privileges, diminished land values and loss of business; their dwellings were marooned in 'an ocean of sludge'.[23]

The application was hastily dismissed but it gave the government pause. Officials realised with considerable alarm that claims for compensation would bankrupt the colony if the government was liable for the sludge nuisance. The secretary for public works, Thomas Balmain, pointed out that there was 'scarcely a river which is not prejudiced or threatened with injury' by sludge.[24] Officials hastily looked for more expedient solutions.

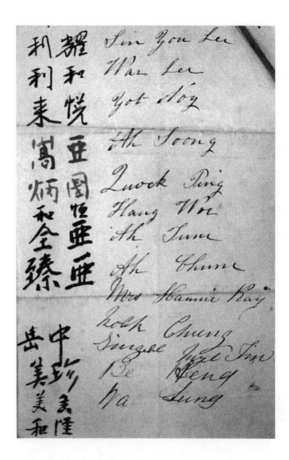

Chinese at Black Lead, petition to the Borough of Creswick

With money from the Public Works Office, helping councils build channels to divert the sludge away from towns looked like a much cheaper solution than paying compensation. During the 1860s and 1870s, Bendigo, Ballarat, Castlemaine, Carisbrook, Chewton, Chiltern, Clunes, Newstead, Maryborough, St Arnaud, Tarnagulla and Tullaroop all received funding, and eventually the farmers at Tarrawingee got a channel too.[25] The government was giving money to anyone who wanted it in a frantic bid to make the problem go away.

The experience in Bendigo was the real catalyst for sludge channels. By the mid-1850s there were 1500 puddling mills on the diggings and the original waterholes were rapidly filling with sludge. Mill operators were encouraged by the mining warden, Joseph Panton, to cut drains for their own sludge, but the ad hoc network that resulted made the problem worse. The sludge overwhelmed Bendigo Creek, and then winter floods spread a thick layer of mud around the shops and houses further down the valley.[26] In 1857 the government spent £7200 on a new main channel and clearing, widening and deepening the old one. Puddlers then petitioned the government for permission to run their sludge into what had been intended as a stormwater channel, and inevitably it filled up too.[27] Heavy rain and flooding in April 1858 brought further slimy misery, with public meetings demanding answers and complaints reaching fever pitch.[28] All winter both the sludge and the public outrage worsened. With pressure on the government mounting, it was decided that a royal commission was needed to resolve the problem of Bendigo's sludge (see figure on page 149). Little did anyone know that it would be the first of many.

Bendigo sludge

> REPORT
>
> OF THE
>
> ROYAL COMMISSION,
>
> APPOINTED
>
> TO ENQUIRE INTO THE BEST METHOD
>
> OF
>
> REMOVING THE SLUDGE
>
> FROM
>
> THE GOLD FIELDS;
>
> TOGETHER WITH
>
> PROCEEDINGS OF THE COMMISSION, MINUTES OF EVIDENCE, AND
> APPENDICES.

Royal Sludge Commission report, 1859

The eight appointed commissioners sat almost every day between 30 November 1858 and 10 February 1859 and heard testimony from thirty-four witnesses. The commissioners were a motley, feisty group, chaired by Irish-born James Forester Sullivan, a veteran of the Mexican–American War and an ex-lieutenant of the Louisiana Guards. He had been a successful merchant on the Californian goldfields before trying his luck in Victoria (and was later elected to parliament to serve as the first minister for mines).[29] Thomas Carpenter was a mining engineer; Richard William Larritt planned and surveyed the township of Sandhurst; Edward O'Keefe was a building contractor who also sat on the Bendigo Mining Board. The previous year O'Keefe had got into a punch-up at a Board meeting over his alleged 'underhand influence' – the business he conducted doing sludge drainage was a clear conflict of interest.[30] There were a lot of interests at stake in the sludge problem.

Vigneron Jacques Bladier, James Sandison and other Bendigo residents offered their testimonies on the extent of the damage and their opinions about how to fix it. Sandison observed that several thousand puddling mills sent 132,000 cubic yards of sludge down Bendigo Creek every day.[31] Only a few downstream business owners and representatives of the Roads Board ventured to suggest that miners should be made to stop releasing sludge. Everyone else agreed that was out of the question: 'It could not be done', declared puddler Gilbert Browne, 'and they might as well give up puddling all together'.[32] The cost of constructing dams and the limitations of land on their own leases were prohibitive, and anyway the sludge would not settle properly. When the commissioners submitted their report to parliament there was consensus: puddlers were 'the great producing class of the district' and it was inconceivable that they should be forced to change.[33]

The commission's recommendation was a familiar echo of most early industry attempts to resolve the problem: a channel to carry the sludge away. Indeed, such an outcome was practically guaranteed before evidence had even been heard. The Commission had been appointed in the first place to 'Enquire into the Best Method of Removing the Sludge from the Gold Fields'. No matter what Jacques Bladier claimed about the loss of his prize-winning vineyards he was never going to convince the vested interests of industry and government that resource extraction should be curtailed in any way. The government reluctantly agreed to pay for a sludge channel in Bendigo and soon extended funding to other towns as well.

The Bendigo sludge drain was an elevated timber channel that ran for twenty kilometres along Bendigo Creek from Kangaroo Flat in the south to a swamp near Huntly in the north. Along with extra drainage works, the total projected cost was £19,000 – a lot of money but, as Bendigo residents were keen to point out, only a fraction of the revenue Melbourne had claimed from the mining industry in lease and licence fees, miners' rights and export duties on gold. Work on the planking commenced quickly but money ran out with several miles of the central section incomplete. Further funding was granted and an additional stormwater channel was also dug along the bed of the creek to help alleviate the problem. The timber lining, however, proved a short-sighted solution and the central section was gradually replaced with stone flagging. By the late 1860s the dilapidated timber sections were dismantled, although with the number of puddlers at Bendigo down to 400, the worst of the sludge problem was by this time abating.[34] Today the channel is a massive stone-lined drain that carries Bendigo Creek through the heart of the city; like those on other goldfields towns, it remains a significant feature of civic infrastructure.

Scientists call these interventions 'channelising' – making streams flow within artificial ditches. It's one of the more extreme forms of human attempts to control rivers, and it has terrible outcomes for the natural environment, particularly when the channels are lined with masonry or concrete as they are in goldfields towns: vegetation cannot grow, habitat for fish and insects is destroyed, and the water is disconnected from the context of its bed, banks and floodplain. Nutrients no longer nourish the floodplains and surface water can no longer percolate down through the soil to replenish groundwater supplies. Although local erosion and flooding are controlled, channelising usually has a disastrous effect on the river above and below the channel. Upstream of the channel the river erodes more quickly, and downstream the eroded sediment piles up.[35]

Now that the sludge is gone some communities have mixed feelings about these old drains. People are starting to place a higher value on waterways as natural environments. In some parts of central Victoria, communities are asking whether the sludge channels could be made a little wilder again. In Castlemaine, for example, local groups have spent several decades clearing weeds and planting native vegetation along parts of the original creek channel.[36] Mature eucalypts now shade the banks and deeper pools are forming, providing homes for platypus. Now the community is considering a proposal to reinstate a narrower, deeper, more sinuous stream within the wide sandstone-lined channel of the 1860s sludge drain. Permanent pools and the replanting of native vegetation would help improve habitat and amenity. In Bendigo too, the Dja Dja Wurrung people are already making Bendigo Creek the focus of their efforts to heal country.

Other communities are taking the opposite approach. Creswick suffered considerable damage to homes and businesses in flooding from 2010–2011 and the ongoing psychological impact has been high. As the community fears the risk of further damage, they've extended and enlarged the Creswick Creek flood channel. New levees have been built, existing ones made higher, and the channel has been deepened in places.[37] The Creswick sludge is long gone but the need for flood control remains.

Channelising was the core strategy of the first battle in the war against sludge, and in general it achieved its aim of moving the sludge somewhere else – out of sight and out of mind, to places where there were fewer people to complain. It is those existing complaints, however – those that have survived in the archives – that have made the sludge visible as a part of local history. It lingers in documents from times and places where there were both enough people and sufficient damage to their livelihoods that someone made a fuss. In the 1850s, that fuss was made in goldfields towns and channels were the favoured intervention. Downstream, those affected were scattered farmers and graziers and the few surviving Aboriginal people, so sludge was more likely to go unnoticed, or at least unprotested. By the end of the 1860s, however, land reforms were breaking up the old pastoral leases and people were leaving the goldfields to start farming. More people were living on the land, and the second battle was about to begin.

Laws regulating gold mining in the early 1860s included provisions to restrict sludge, but these were still mostly ignored. Schedules of mining leases issued under the *Mines Amending Act* in 1862, for example, directed miners to dispose of 'detritus, sludge, rubbish or other waste' from their operations so that none of it flowed into waterways, other claims, roads or public or private property.[38] But this was an obscure provision in tiny print buried deep within a long, complex legal document. Few observed the restrictions in practice – especially when nobody else did. With enforcement left to the Mines Inspectors, little action was taken.

In 1865 the government consolidated the piecemeal laws and provisions that had accumulated around mining since 1851. The 1865 *Mining Statute* was one of the most important pieces of mining legislation ever passed in Australia, providing the foundation for much of mining law throughout the Australian colonies and New Zealand.[39] What it said about sludge was thus potentially very significant. The *Statute* sounds promising – under section seventy-one mining boards were given power to enact by-laws to prevent sludge accumulating, appoint sludge inspectors and collect fines for breaches. There were also potentially fines for polluting reservoirs. Essentially though what the *Statute* said was more of the same – managing sludge was left to the same group of elected representatives that oversaw all other aspects of mining. Continued self-regulation meant that Mining Boards took some interest building sludge drains but little else. The main response was still to move the sludge somewhere else rather than prevent it at the source.

The real problems started when the drought of 1865–1866 broke with heavy rains across the colony. During the dry years mining waste had accumulated in the upper sections of river catchments. Stormwater now shifted the accumulated sludge downstream, dumping it on the new farms spreading across the northern plains and south of the Dividing Range. The little town of Inverleigh started to get Ballarat's sludge in the spring floods of 1868. More came down the Leigh River in 1869 and 1872.[40] The Inverleigh correspondent for *The Ballarat Star* complained that 'The sludge brought down the Leigh from Ballarat has covered the low-lying lands of Captain Berthon's property a foot deep, and also enveloped his brother's property and others near the river' (see picture section).[41] Berthon himself claimed that 'what were once rich alluvial flats are now nothing but an impervious mass of sludge. The waters are polluted and unfit to drink, and the natural channel of the stream being blocked up, causes a flood to be doubly dangerous'.[42] He called for government action; while his neighbour, George Russell of Golfhill station, built his own levees to keep the sludge off.

The sludge problem made it all the way to Victoria's Supreme Court in 1869, when a Mr Campbell took out an injunction against a party of Chinese miners headed by Ah Chong. Campbell owned land at the Bald Hills near Creswick and claimed the miners were sending sludge from their works down a creek that ran through his property. The court found in Campbell's favour, deciding that it 'is incumbent on persons mining on Crown lands under miner's right to use a stream, flowing past the ground on which they are mining, in such a way as not to injure the land or the water of those below them'.[43]

Campbell may have won an expensive legal battle, but the larger war against sludge continued. Around Tarrawingee, sludge started creeping down in the floods of 1867; it returned in 1869 and got worse in 1870 when autumn rains inundated the district in April and spring rains flooded it again in September and October. The local paper reported that 'farms were flooded on every side and as usual an enormous quantity of sand and sludge was deposited on the cultivated land'.[44] Farmers did all they could. Like residents of the towns, landowners saw the solution in a channel that would move sludge on elsewhere. From 1870, editorials and letters in the *Ovens and Murray Advertiser* called for a sludge channel.[45] Farmers held local meetings and sent deputations from the North Ovens Shire to Melbourne to lobby the minister for public works, the treasurer and the premier. Local MPs made speeches in parliament and tried to convince the minister for railways to take an interest in damage to the rail line.[46] Three surveys had been carried out by the end of 1875, but there was still no channel. In desperation, some Tarrawingee farmers started to dig their own cuts. The next winter two-year-old Willie Whittle wandered into the sludge and was saved only by good fortune.[47] Severe flooding in the winter and spring of 1877 prompted nearly 100 local men to gather repeatedly in a series of public meetings, and finally, in the autumn of 1879, construction of the channel began.[48]

At Inverleigh, after the floods, the locals got a parliamentary inquiry almost immediately, but its results showed that even political influence was not enough to get action. Major landowners included wealthy pastoralists with far more political influence than the small farmers at Tarrawingee.[49] They got the matter raised in parliament and so the Ballarat district surveyor, Henry Morres, was sent to inspect the region in the spring of 1872. The outcome of Morres' inquiry, however, was further inaction. Morres noted that 'much damage has been done (between Shelford and Inverleigh as well as) at Ballarat and in various places along the whole course (of the Yarrowee/Leigh River)'. He admitted that 'the chief cause of damage is because the miners, with but a few exceptions, purposely run their sludge into the nearest watercourse', and that it had been going on for years and could be expected to continue.[50] Morres' conclusion seemed to be that the habits of miners had congealed into an unassailable status quo; 'the practice of running sludge into the watercourses commenced simultaneously with the discovery of gold, and has been the practice ever since', despite the fact that there was no law that expressly allowed them to do it. 'As regards the farmer,' Morres wrote,

> he is injured, but by whom? Not by one person or company only, but by the whole mining community who have custom on their side; he therefore hesitates before commencing an action at law against any person or company, as he would, in all probability, have to sustain an expensive lawsuit with but little chance of obtaining satisfactory redress.[51]

To take action, landowners suffering injury needed to trace the damage back to a single source – a person or company who could be held responsible in court. At Inverleigh and Shelford the source was the entire community of Ballarat miners, which meant little prospect of winning.

Morres thought the remedy might be drainage legislation, similar to that employed in England, which would make it illegal to allow mining waste to flow into watercourses. He offered wording and clauses that he thought might do the job, and the local member, pastoralist William Robertson, seized on Morres' report to demand a bill be put before the Legislative Assembly. The minister of lands said he would think about it.[52] Mining was simply too powerful.

Occasionally injunctions were successfully imposed. In the 1860s Justice Robert Molesworth granted several injunctions restraining miners from discharging sludge onto the lands of other companies and nearby watercourses. Molesworth was chief judge of Victoria's

Court of Mines and for many years had a profound influence on the development of Australian mining law.[53] In 1867 at Ballarat, for example, the Prince of Wales Company was restrained from allowing sludge to flow over the land of the adjoining Bonshaw Freehold Company and into Winter's Creek.[54] The polluted water was unfit for use in the boilers and for washing out the gold. The company argued that the sludge came from other companies as well, but Molesworth accepted the evidence that this company was the specific source of the pollution. This set an important legal benchmark, but it did little to stop the sludge problem from continuing everywhere else.

As Morres was preparing his report to parliament, barrister John Atkins was also thinking about the legalities of sludge, among other mining matters. In 1871 he published a lengthy treatise interpreting the 1865 *Mining Statute* in which he reviewed the results of court cases and gave his opinion of where the law now stood. One of the problems was a lacuna in the legislation: it didn't deal with sludge in sufficient detail. Section 34 prohibited miners from allowing sludge to pollute reservoirs but this fell far short of stopping other damage to rivers, businesses, roads, homes and farmland. There were only two relevant cases to draw on, but from these legal precedents Atkins concluded that 'A person has no right to discharge the water used by him loaded with sludge, on the land, or into the water, of another ... and may be restrained by injunction from doing so'.[55]

Landowners seized on these precedents. Inverleigh may not have wanted to bring a legal case against the whole of Ballarat, but elsewhere landowners could and did identify specific sludge-producing mines. In 1875, injunctions were granted against companies at Mount Egerton and Maryborough. *The Maryborough and Dunolly Advertiser* responded with outrage: 'It appears that any person, no matter who, can at any time ruin a mine in this way. The thing seems very laughable and absurd, but law is law'.[56] In 1881, landowners around Clunes were granted injunctions against two Creswick companies, the Australasian and the Madam Berry. These were deep lead mines that worked ancient riverbeds buried beneath later lava flows. At the time, they were some of the richest goldmines in the world. The Madam Berry mine alone produced more than 370,000 ounces (twelve tons) of gold between 1881 and 1895.[57] Injunctions were also threatened against four other companies in the district: the Loughlin, Ristori, Lone Hand and Dykes Freehold.[58] If these injunctions succeeded and the mines closed, more than 1300 miners would be thrown out of work, not to mention the losses to rich investors in Melbourne.

When the colony's biggest mines were targeted, parliamentarians started to pay attention. Mine owners hurried to Melbourne to lobby the government and the Sludge Drainage Bill was quickly drafted, the first time that parliament was formally asked to consider the issue.[59] The specific aim of the bill, however, was not to limit sludge; it was to send the sludge into waterways more efficiently. Creswick miners wanted to build their sludge drains without interference from the landowners through whose land the channels passed, and that was what the bill promised. Its principal clause granted access to private land so that channels could be built. There would be no more danger of injunctions, and mining would continue unhindered.

Parliamentarians from agricultural districts immediately objected to the threat to private property rights embedded in the bill. As the debates progressed, members from mining districts became increasingly concerned too. Their constituents became more and more alarmed as they followed proceedings reported in daily newspapers. They did not hesitate to express their views, and politicians were bombarded with letters, telegrams and petitions.

The problem for miners was an unexpected clause inserted by Robert Burrowes, the minister for mines and member for Sandhurst during the drafting of the bill. As he introduced the bill to the House, he explained that 'Beneficial as gold mining was to the community, there was no doubt the sludge from gold mines was beginning to do considerable harm to the surrounding country, and it would be well to restrict that harm to the narrowest possible limits'.[60] He thus added a clause that enabled mining wardens to compel miners to build dams that would prevent sludge from entering creeks and rivers. Miners would have to stack their sludge rather than flush it away. With this one clause a significant new dimension was added to the issue. Burrowes had unintentionally directed attention to something that the mining community would have preferred to leave unexamined.

Furious debate ensued as miners realised what had happened. Members from mining districts explained to Parliament in no uncertain terms that this was not what had been intended at all.[61] Major William C. Smith, representing Ballarat West, thought keeping sludge out of streams unnecessary. 'Had not hundreds of thousands of tons of sludge been conveyed into the Yarrowee Creek at Ballarat, and what damage had resulted?' he asked. 'It would be 'monstrous', he asserted, 'to stop mining proceedings simply because half-a-dozen small farmers were affected'.[62] George Kerferd from Beechworth said it would 'prohibit mining entirely' in his region – as the land was so mountainous, there was no alternative but to send the sludge into streams. James Brown Patterson worried that farmers on the Loddon would be able to put an end to mining at Castlemaine, while James Henry Wheeler pointed out that in Creswick miners worked in the actual beds of the creeks. John Woods from Stawell was concerned it would end the pumping of contaminated water from quartz mines. Representing Maldon, Joshua McIntyre stated that 'If the measure passed in anything approaching its present form, it would stop gold digging throughout the colony. It meant the absolute prohibition of gold mining, both quartz and alluvial.' Mining interests were getting more and more frantic as the debates gave members from non-mining areas a rapid education in what was happening to the colony's rivers. Robert Murray Smith from Boroondara in Melbourne's eastern suburbs summed up a view that was ultimately shared by many: '[regarding] the relations of mining companies and mining generally to the water supply of the country ... its importance grew as it was being considered.'[63]

William Lawrence Zincke had a more nuanced view. Zincke was the newly elected member for Ovens and had decades of experience in mining law to draw on. He was born in Jamaica and educated as a lawyer in England before migrating to Australia in 1848, and later setting up a legal practice in Beechworth. There he was a much sought-after solicitor, drumming up plenty of business in the mining community, and in 1880 he had acted for Ned Kelly sympathisers after the siege at Glenrowan. In Zincke's view the riparian rights of landowners were semi-extinguished if the water had already been diverted for mining before the land was occupied. The riparian proprietor 'was entitled to the same volume of water that existed in the stream when he purchased his land but, if the water was already polluted, he had no right to complain if it continued to be polluted'. Only if the land was purchased prior to mining water diversions was the landowner a clear riparian proprietor, entitled to 'water undiminished in quantity, but also undefiled in quality'.[64]

Working out the rights of miners versus farmers and the relative importance of rivers, mines and farmland was a messy legal and political business. All these arguments about compensation and riparian rights were discharged against a backdrop of wider parliamentary and public debates over the relative values of leasehold and freehold land, and which better served the cause of social and economic development. What came first, private rights or the public interest? The *Lands Compensation Statute* 1869, for example, had little to say about mining, except that holders of miner's rights and leases were not entitled to compensation if they were to be deprived of the land by the government for public purposes.[65] The situation was complicated in May 1881 when the government formalised its long-standing custom of reserving strips of unalienated land, known as frontages, along 280 of the colony's rivers and streams, which meant no new riparian land could be acquired. This was part of a gradual assertion of official control over water resources that had begun years earlier with the issuing of licences and permits to miners diverting water. [66]

MPs from mining districts recoiled in horror from the sludge control provisions of Burrowes' bill and rejected it en masse. Petitions against the bill were presented from miners in Walhalla and Sandhurst. Major Smith of Ballarat West declared that 'there was not a single gold-fields member willing to accept the measure ... They would far rather the miners were left as they were'.[67] With their new knowledge about the condition of the colony's rivers, MPs from metropolitan and agricultural districts supported the bill enthusiastically and it passed the lower house. In the Legislative Council, however, mining interests held more sway. Mining entrepreneur John Alston Wallace attempted to amend the bill in favour of his Beechworth business partners. Sir Charles Sladen from Geelong, among other MPs from agricultural districts, argued that the protection for rivers did not go far enough. In the end the issue was too complex to resolve and in a narrow vote the bill was abandoned. The Creswick miners quietly paid compensation to the landowners, got their injunctions lifted, mining continued, and the sludge continued to fill the rivers.

Reading transcripts of the debates in Hansard, one begins to get a good sense of what was at stake. Members repeatedly claiming that regulating sludge would mean the end of mining tended to cite the large numbers of people who would be left out of work. In 1881 there were still more than 38,000 men directly employed in mining around the colony, out of a total Victorian population of about 900,000 people.[68] With the bill, mining interests were trying to achieve greater security for their industry and protections similar to those in place in New Zealand. Under New Zealand's *Gold Fields Amendment Act* 1875, rivers could formally be declared 'sludge drains' and any person whose rights were injured or affected was entitled to compensation.[69] The 1881 debates demonstrated conclusively that such an approach could not work in Victoria. There were too many competing interests and not enough water to go around. Some landowners along rivers still held riparian rights to water and rights to compensation if their lands were damaged. Considerable public money had recently been invested in water supply systems for goldfields towns as well, including the Coliban system that supplied Bendigo and Castlemaine. Only recently completed, the Coliban was massive, expensive, controversial, and at the forefront of politicians' minds. And even Coliban water could not be completely protected. Parts of the system's catchment boundary and its tributies were downstream from mining activities and so any measures to keep its water clean and clear would inevitably impede mining. The premier, Sir Bryan O'Loghlen, pointed out that it would be contradictory to allow water to be contaminated in the Sludge Drainage Bill while trying to preserve it in the Water Conservation Bill, which was also before the House.[70]

A further complication was the enormous diversity in mining conditions across the colony. What worked for underground alluvial mines on the flat country around Creswick was anathema to hill-sluicers in the mountains of the north-east and to quartz miners at Stawell, Bendigo and Walhalla, where the topography or the chemistry of ore bodies created a different set of problems. New Zealand's simple expedient of legislating rivers as sludge drains was achievable because mines there at this time were mainly surface alluvial, located in mountainous districts with large, snow-fed rivers, and in locations where land was still leasehold. The rivers declared to be sludge drains carried huge flows of fast water and were very efficient at disposing of mining waste. The Victorian situation was more complex with a diversity of mines and rivers that lacked snowmelt and carried far less water. Any attempt to control sludge while allowing mining to continue would have to be flexible enough to encompass ground sluicing and underground workings, quartz mines and alluvial deep leads, and mining in cramped mountain valleys as well as on wide plains. In 1881 it proved impossible to either adjudicate the dispute or even achieve consensus among miners.

The debates in parliament also shed light on underlying assumptions about how mining operated and its role in the colony. Mining members could not agree on how to control sludge – or if such control was even necessary – but they all shared the view that miners had a legitimate right to send waste into rivers. John Gavan Duffy argued that 'the question of protecting streams against sludge was not a matter of urgency. The running of sludge into streams had been going on for the last twenty-five or thirty years, and no complaints had arisen about it.' George Kerferd agreed: 'for at least the last quarter of a century the practice of the miners ... had been to let their sludge run into the local water-courses'. Likewise Richard Richardson, who said that 'to prohibit them from doing so would take away something which they had enjoyed for the last twenty-five or thirty years'. Once again miners felt that custom and practice were on their side, and this was just the way things were done in Victoria. Some went so far as to claim that the sludge was even beneficial: Hugh McColl maintained that 'the tailings of Bendigo had enriched every acre of land on which they had been deposited'.[71] McColl must not have seen

Jacques Bladier's vineyard at Epsom or he would not have spoken so glowingly about the apparent benefits of Bendigo's sludge.

When members spoke persuasively about time-honoured rights to discharge sludge they were cavalier in their claims that there had been no complaints. Major Smith airily described hundreds of thousands of tons of sludge that had gone down the Yarrowee without damage, conveniently forgetting George Russell and the Berthons and the recent inquiries at Inverleigh and Shelford. The long struggles of Tarrawingee farmers were likewise overlooked by William Zincke. He asserted that landowners downstream from Beechworth had 'lost their rights by having slept over them for twenty years and made no attempt to assert them during that period'. In other words: use it or lose it. One of his colleagues in the Legislative Council, Nicholas Fitzgerald, went so far as to claim that in twenty-five years of mining in Beechworth 'there has not been the slightest complaint by any riparian proprietor of damage done by sludge'.[72]

After the collapse of the 1881 bill the sludge continued to flow. Up at Tarrawingee, the drain that had been started in 1879 was, six years later, still not solving the problem. The plan was to channelise and straighten Hodgson Creek so that the sludge would bypass the natural meanders of the waterway, and flow quickly through to the Ovens River. But the works were incomplete; two councils were involved, but with insufficient money to finish the job. The Shire of North Ovens started digging the channel north from the Ovens River and then east to the Tarrawingee railway station. Beechworth Shire built their section heading west from the hills. The upstream section at Beechworth, however, was designed to carry a much larger volume of mining waste and floodwater. The North Ovens Shire resented having to build the channel at all, and so dug it on a smaller scale. By 1885, construction from each end approached the railway station but stopped with a gap of thirty-seven chains (about

700 metres) between them. Money had run out and the truculent councils couldn't agree whose job it was to finish the channel. As President of North Ovens Shire, Hopton Nolan thought it was time the government stepped in and paid, even though the money had come from Melbourne in the first place. More letters to the newspaper, more community meetings, and more lobbying of politicians occurred, and the sludge from mining in the hills kept spreading over farms on the flats. When members of the Board Appointed to Inquire into the Sludge Question (the Sludge Board) arrived in Wangaratta in December 1886, feelings were running high on both sides of the gap.

The Sludge Inquiry of 1886 was yet another attempt by the government to solve a problem that only seemed to be getting worse. After the injunctions episode in the early 1880s the sludge question may have left the front pages of the Melbourne newspapers but it hadn't gone away. More activity around Creswick caused the issue to flare up again in 1885.[73] New mining companies had formed to exploit the wealth of the deep leads and after years of digging shafts they were about to start raising washdirt and generating sludge. By now they knew that they would need the cooperation of local farmers. Mine managers wanted to drain their waste into Birch Creek, but a weir on that stream, built by the Port Phillip Company in 1866 (ironically, to supply their own mining operations), provided water for the town of Clunes. The deep lead miners would thus need to divert their waste to a point below the weir to avoid contaminating the Clunes water supply. They needed to acquire land from the farmers to build drainage channels and they needed the farmers' permission to discharge the sludge.

Farmers and graziers had also acquired knowledge by this time. They knew to be wary of mine operators, and they also knew the likely value of the mines. In February 1885, as the mines prepared to begin processing, the landowners acted. They sent a deputation to Melbourne to see the minister for mines and water supply and alert him to the problem. They approached the shire councils of Clunes, Talbot and Creswick to join them in lobbying. After meeting collectively with representatives of the ten new mines, the landowners presented their demands. They would grant access, but they wanted thousands of pounds in compensation for the drainage, in addition to rent for use of the property and a share in royalties from the mines. These were breathtaking demands.

The mine managers were outraged. By October they had become frantic, and their representatives gathered in a mass meeting at Craig's Hotel in Ballarat and resolved to act. They knew that other mines around the colony were facing similar problems, and they called on members of the Mine Owners' Association and the Miners' Association to meet at Trades Hall in Melbourne. The Melbourne meeting got the attention of politicians, and eleven sitting members from goldfields districts took the cause directly to the premier, the treasurer and the attorney-general. The miners wanted 'equitable settlement of the sludge question', which for them meant being able to drain their waste without interference. In November 1885, the minister of mines and water supply and the secretary of his department arranged to visit the deep lead mines around Smeaton; he promptly announced that a bill to resolve the issue would be put to parliament before Christmas. In spite of the problems in 1881 they were willing to have another try at legislating a solution.

The new bill was debated in December 1885, on the last two sitting days of the year, just before parliament broke up for Christmas. This time the attorney-general drafting the bill was the soon-to-retire member for Ovens, George Briscoe Kerferd. With this bill there would be no repeat of the suggestion that mines should be prevented from discharging sludge into rivers – the focus now was on how sludge would be managed with minimal interference from land-owners. Its principal objectives were to insist that sludge be held in settling ponds before being sent into channels, to create local sludge boards that would build and maintain drainage channels in each district, and to allow mines to continue discharging waste into rivers if they had been doing so for the previous five years. The debates revisited the issues canvassed in 1881. MPs from other mining districts were anxious about the effect on quartz mines, while non-mining members were concerned about the ongoing damage to rivers.

The eleventh-hour, two-day process used to get the bill through was protested heartily by some sitting members. Many had already left for Christmas, and those few remaining were finally worn down. Representatives from mining districts succeeded in restricting the provisions so that they applied only to deep lead mines in the Creswick district. At eleven p.m. on 18 December the House passed a much more specific bill: *An Act to provide for the Disposal of Sludge from Alluvial Mines in Creswick* (1885). It was not actually about reducing sludge, however. As with earlier initiatives to build sludge channels in goldfields towns, yet again, the real purpose of the Act was to move the sludge somewhere else. It was legislation drafted to protect the miners, not to protect the rivers. Those who sought to protect the rivers nevertheless secured one small but significant victory: the clause that would have made it legal to continue discharging sludge if had been done for five years already was struck out.

Instead of laws with teeth, those concerned about the health of rivers and water now got another inquiry. A Board 'to inquire into and report on the extent of the injury caused by Sludge from Mines, and the best means of preventing or mitigating future damage' was gazetted in early August 1886.[74] It was chaired by civil engineer Robert Shakespear, who had been involved in the construction of the

Coliban water supply system, and included Sandhurst mining surveyor Arthur F. Walker and R.G. Ford from the Public Department
Works. These men were diligent and took their role seriously.[75] By
the time they presented their report seven months later they had visited seventeen communities across the colony, from Mitta Mitta in
the far north-east to Newbridge, Carisbrook and Maryborough on
the Loddon plains. They inspected twenty-nine creeks and rivers,
took dozens of water samples and visited seventy-eight quartz and
deep lead mines as well as sluicing claims in ten mining districts.
They interviewed 163 witnesses, including fifty-four miners, seventy-
five farmers and thirty public officials. Their report was published
in 1887 and was nothing if not thorough; it remains a landmark in
describing the extent of the sludge problem after thirty years of gold
mining in Victoria.

While parliamentary debates had previously been steered by
mining interests, the Sludge Board found itself overwhelmed by the
compelling evidence of damage. The issue, as it saw it, was not draining the mines but preventing sludge damage in the first place. This
was a problem both widespread and serious, as its catalogue of 'evils
arising' insisted. It reported:

> The injuries already inflicted, and which, unfortunately, in
> many cases, cannot be cured, consist in the filling up of the
> large clear waterholes in the creeks and rivers (used for stock
> and domestic purposes), the silting up of the river beds, causing
> the sludge to overflow on the adjacent lands, to the destruction
> of vegetation and fruit trees; the liability of horses and cattle
> going to water in the creeks being bogged in the slum and per-
> ishing there, or contracting disease by drinking the muddy and
> often mineralized water; the deterioration in the quality of the
> wool of sheep depasturing in the vicinity of the silted-up

streams, by reason of the sand being blown into the fleeces by the wind, and the depreciation in the value of the lands affected. These, and the destruction of roads and bridges, are some of the evils arising.[76]

The puddlers in Bendigo were long gone, but sludge was still being produced there by sluicers and quartz mines. The sludge channel constructed in the 1860s was now full to the level of the bridges, and Bendigo Creek was obliterated at Huntly. Around Wangaratta the damage to the Ovens River was deemed 'irreparable' below the junction with Hodgson Creek, and both Hodgson and Reedy Creek were terribly affected. The lower Loddon River was badly filled as far as Eddington, as was the Leigh River between Shelford and Inverleigh.[77] The Board found that quartz mines were just as damaging as hill sluicers. Quartz crushing in the mines at Clunes and Maryborough was sending thousands of cubic metres of tailings into the Creswick and Deep creeks around Carisbrook and down to Eddington. Sludge filling the Leigh valley at Captain Berthon's farm came from quartz crushing at Ballarat.[78]

On the brighter side, the Board was heartened by the mines where it was evident sludge was being managed successfully, particularly around Bendigo.[79] Quartz mines along Myers Creek were considered exemplary in the way they stacked their tailings, with the Board expressing admiration for the clear water in Myers Creek. At White Hills the sludge from puddling had filled in the shafts left by old workings, and the land now supported good crops and orchards. Tailings had been used to good effect in Eaglehawk, where old workings had been filled in and converted to public gardens. At Yackandandah, sludge was being diverted into workings along Yackandandah Creek, and the Board predicted that valuable land would soon be created there too.[80]

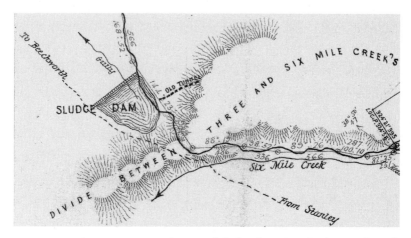

*Proposed sludge dam, Friedrich Kassebaum's
application for a water-right licence, 1881*

Some of the hill sluicers around Beechworth were being forced by
necessity into using settling dams, and their efforts provided another
example of what was possible (see figure above). Companies in the
district frequently reused water from neighbours as their source of
supply, a system the Board examined in some detail.[81] Upstream
companies were required to pass on waste water in a useable state,
which meant that settling dams below one operation acted as supply dams for the next company along the valley. The Excelsior
Company on Spring Creek, for example, stacked tailings from the
Rocky Mountain Extended and the claims run by Donald Fletcher
and used it as their water supply. As soon as sludge flow was slowed
by the dam the sand and gravel would settle, leaving clear water that
could be pumped out and used again.[82] John Pund and his neighbours on Three Mile Creek operated in a similar manner, competing
and cooperating at the same time. Pund reused tail water sent down
from William Telford's United Sluicing Company, then settled his
own sludge in a disused part of his claim before discharging it downstream to James Gillies, who stacked his debris behind a lightweight

sludge dam of brush and branches. Philip Busch used water from three upstream parties (Chambers, O'Neill and Welsh) and sent it down the valley to a Chinese party headed by Ah Ping. Ah Ping also got water from Lawrence Murphy, who worked above him on One Mile Creek.[83]

These collaborations aside, the Board was unequivocal in its grave warnings for the future. In particular, it predicated a whole new set of problems along the Loddon as water from the Coliban System became available. In 1886 when the Board visited Castlemaine, the water had reached as far as Fryerstown, and additional channels were planned that would eventually deliver water across the ranges to Vaughan by 1894.[84] Cheap and abundant water was already transforming the mining industry in that district. As we saw in Chapter 2, innovative mine managers like John Ebbott at the Garfield Company realised that water power was now less expensive than wood-fuelled boilers and they were converting their power supply from steam to waterwheels. Firewood for boilers cost five to six shillings per ton, with the added cost of an engine driver, while a penny would buy 1000 gallons of reliable Coliban water.[85] What concerned the Board was the increased capacity of such mines to generate sludge. Ebbott admitted that his company was also using the Coliban water to wash tailings into the nearest creek rather than stacking them at the mine as they had done before.

Quartz tailings were likely to be the least of the problems, though. The lower reaches of the Loddon River were already struggling under the sludge load carried down by the tributaries that drained the mines at Daylesford, Maldon, Creswick, Smeaton, Clunes, Maryborough and Dunolly. So far, the upper tributaries around Castlemaine had remained relatively clear of mining debris. Lack of water supply in the region had kept sluicing to a minimum and, as yet, not much sludge had been released. But that was about to change. The Sludge

Board heard from sluicers already using Coliban water on Forest Creek around Chewton and spoke to others along Barkers and Fryers creeks who were planning to use more Coliban water in their operations as well.[86] The Board warned: 'On completion of the Coliban races ... more extended sluicing operations are anticipated ... all tending to fill and discolour the Loddon'.[87] Landowners at Yapeen were already suffering from sludge in Campbells Creek and those at Eddington knew the Loddon was less muddy when the sluicers on Fryers Creek stopped working for a day or two.

The Board reserved its highest note of alarm for the looming threat of new technologies. Californian miners had continued to tinker with the equipment used for hydraulic sluicing, and by the 1870s they had perfected the giant nozzle.[88] These were powerful water cannons with nozzles up to nine inches in diameter, capable of bringing down riverbanks over 120 feet high. The Board heard that three of these monsters were already in use in Victoria, with more on their way, particularly in the north-east where plentiful water and the steep terrain of the Mitta Mitta valley made their use ideal. A single giant nozzle could shift a ton of earth every minute. The Board calculated the effect if the number at work increased to twelve, as was planned; it estimated that in one year up to two million cubic yards of sediment could be produced, 'sufficient to fill a stream 105 feet wide to a depth of three feet for a distance of thirty-two miles'.[89] Of the giant nozzle it stated, categorically, that 'this is the proper time to regulate its further use in Victoria, before further damage is done or further vested interests can accrue'. It even corresponded with officials in California to gather more information about the nozzles' effects there. Its report highlighted the recent Sawyer decision (1884) in the US Circuit Court that had banned hydraulic mining in California, unless it could be carried out without damage to other interests.[90]

It's clear from their report that Board members were genuinely shocked by what they found, despite their own personal experience of the goldfields. 'The damage caused by mining sludge throughout the colony is of a far more serious nature than was generally supposed', they wrote, and 'the injury is likely to be greatly increased in the future ... unless remedial measures are immediately taken'.[91] They did all they could to awaken their peers to the problem, publishing a long and detailed report with transcripts of witness testimonies. They had uncovered the previously unpublished statements made by witnesses at the Bendigo sludge enquiries twenty years earlier and they insisted that these be made public too.[92]

Unlike the 1858–1861 Bendigo inquiries, the Sludge Board recognised that farmers had legitimate grievances. Mining was doing momentous damage to the wider interests of the colony. In several instances it calculated the quantity of sediment produced and compared it to the value of gold recovered and the number of men employed. In all cases of sluicing the results were clear. Sluicing was, it concluded, 'productive, in a great many cases, [but] of more injury than gain'.[93] It cited a ground sluicing operation at Chewton that generated 300,000 cubic yards of sludge for scarcely enough gold to pay the wages of the four men employed. At Mitta Mitta a hydraulic sluicing claim produced 660 cubic yards of sludge for each ounce of gold retrieved. The Board thought the mining industry ought to act more responsibly: 'where the convenience of others is at stake, and not alone their convenience but their property and means of livelihood, the mine owners should show some consideration ... it is evident that the present mode of disposing of the detritus from quartz mines cannot, in fairness or justice, be allowed to continue'.[94]

The Board saw an urgent need to put an end to sludge, and yet, together with many of its witnesses, it remained reluctant to hamper the mining industry.[95] Farmers at Carisbrook, so badly affected

173

by sludge from Clunes, stated nonetheless that 'we are fully sensible that the industry of gold mining should be fostered and encouraged'. The Sludge Board included the Carisbrook petition to the minister of lands in its own report. Here the petitioners felt their request was modest and not unreasonable: 'your memorialists ... humbly submit that the boon they solicit is but that one of the good gifts of Providence may reach their homesteads in its purity, and free from injurious polluting influences placed in it by the gold-mining companies'.[96] These farmers were not alone in their acknowledgement of the prior rights of miners. Those around Clunes and Smeaton agreed that they did not expect water to be perfectly clear; their stock could and did drink water that was slightly discoloured. The Board used this admission to make the case for action by the miners: 'Some of the agriculturalists recognise the fact that it would be injurious to the community to press objections to discolouration of the water by mining, and agree that, were the mining companies to adopt measures to ensure the best method of stacking the sludge, there would not be any serious ground for complaint on the part of the farming interest.'[97]

In the end the report was conciliatory – 'the subject was felt by all to be ... a difficult and delicate one'.[98] The Board urged compromise and negotiation: 'The difficulties can apparently be overcome without serious loss to the mining community if, on the one hand, mine-owners will consider the injury that a want of care in stacking the debris produces ... and if, on the other hand, the land-holders will see the necessity of leaving the mining interest as little trammelled as possible.'[99] The report took pains to identify reasonable solutions, all of which drew from the examples of some 'best-practice' companies in certain places.[100] Among its proposals were the following:

- On leases where spare land was available, particularly in the case of quartz mines, tailings should be kept in settling dams.

- Where leases were too small other solutions would need to be found, including possible arrangements with neighbouring mines or the diversion of sludge into old workings.
- The government needed to take more responsibility for ensuring that leases were large enough for tailings, and the Board criticised previous policies that allowed land immediately surrounding mines to be sold.

These were all viable approaches for the quartz and deep lead industries, where the economic value of gold and the employment it stimulated were considerable. The sluicing industry was another matter. Economically it didn't make sense to destroy so much land for so little profit. The only solutions were:

- Much more stringent control of leases so that sluicing was kept away from riverbanks;
- Claims only approved if there was sufficient ground for storing the waste; and
- Giant nozzles forbidden entirely.

The Board was so concerned that it even drafted a bill to be considered by parliament 'to Provide for the Drainage of Sludge from Mines'.[101] The draft *Sludge Act* set out a number of interventions. The inspector of mines in each district was to be given the added responsibility of becoming a sludge inspector with the power to enter and inspect mines to examine arrangements for settling sludge, respond to complaints of injury, and assess new lease applications to determine if appropriate arrangements for sludge had been or could be made. All mines and sluicing operations were to make provisions for settling sludge, and there were to be penalties for allowing it to enter onto land or into any kind of waterway that was not a sludge channel. The Mines Department would be given the power to make sludge channels and take responsibility for maintaining the sludge dams of abandoned mines. Where it was found impossible to mine without incurring damage from sludge then no mining should be permitted on that land. Finally, there was to be compensation paid in those cases where existing mining leases would have to be discontinued under the new Act.

It was a well-crafted bill and would have done much to solve the sludge problem, but it was not to be. Like many government inquiries before and since, the Sludge Board had already served its purpose. Neither the report nor any of its recommendations were ever debated in parliament. It may be that the report, despite its conciliatory tone, was too unsympathetic to mining. It is just as likely it was the victim of unfortunate political timing. By 1886 there had been several changes of government, and Sir Bryan O'Loghlen's ministry, which had introduced the two earlier sludge bills, was replaced by a new ministry led by Duncan Gillies. For the next few years, when questions about the status of the report were posed to the House by members from districts affected by sludge they were brushed off. Alfred Deakin, serving as chief secretary as well as minister for water supply, focused his energies on establishing irrigation agriculture.

Most of the questions that *were* raised about sludge during this time were posed by Henry Parfitt, member for Wangaratta and Rutherglen. Parfitt was a Tarrawingee man, a neighbour and ally of Hopton Nolan. He served alongside his friend on the North Ovens Shire Council for more than twenty years and was a driving force behind the Tarrawingee Sludge Channel, defending local interests against 'hostile' governments in Melbourne.[102] Parfitt used his position to keep the sludge issue before the House, but when he left parliament in 1892 the questions ended and the report was forgotten.

By this stage the economy had collapsed and the colony was entering a major depression. Mining was on the wane and annual gold production was down to 600,000 ounces for the entire colony, the smallest yield since 1851.[103] The prevailing political imperative was to do as much as possible to help the industry, not to hinder it. The amount of sludge produced was also declining – the richest deposits had been worked out, mines were closing and workers were leaving the industry for easier jobs elsewhere. By the time the 1891 Royal Commission on Gold Mining asked 402 witnesses what was wrong with the mining industry, barely a handful even mentioned sludge. By then, the deep lead mines at Creswick that had been the richest in the world a decade earlier had seen their best days. The Madam Berry, at the centre of the 1881 injunction battles, was on its way to its closure in 1894, while the other mine involved, the Australasian, had been closed since 1888.[104] The Port Phillip Company at Clunes that had plagued farmers at Carisbrook with its quartz tailings closed the same year after thirty-one years of operation.[105] The big alluvial fields in the north-east at Beechworth and along the valleys of the Ovens, Kiewa and Mitta Mitta rivers were still active but had been shrinking since the 1880s. When the embattled farmers at Tarrawingee finally got their completed sludge channel in 1888 they had good reasons to expect that their sludge troubles would soon be over. Two years later, with just over 2000 miners still at work in the whole of north-east Victoria in 1890, they most likely expected the 'nuisance' would decline of its own accord.[106] In fact, the problems were about to get worse.

6

TURNING THE TIDE

JOHN BOWSER WAS DETERMINED THAT THE TIME HAD come to stop polluting the rivers. He could see all too clearly the state the Ovens was in, and it wasn't good. Its water was constantly muddy, and its deep waterholes were filling up.[1] Sludge was flooding rivers again, and in Bowser's view it was a clear injustice that people in one industry, mining, could profit at the expense of others – the downstream farmers and townspeople who depended on the river for their water.

Bowser had fought injustice all his life. The son of an Indian Army veteran, he'd witnessed intimately how such systemic unfairness could ruin lives. As part of a royal commission into the living conditions of crofters in the Scottish Highlands, he saw how discrimination in agreements between large landholders and their tenants created rural poverty. Now he was a businessman and a journalist in Wangaratta, and the owner of a small farm on the Ovens at Oxley, across the river from Tarrawingee. He'd arrived in to Victoria during the mid-1880s, bought a half share in the *Wangaratta Chronicle* and helped establish the local library as well as clubs for shooting and tennis. His reputation as a staunch supporter of rural affairs saw him elected as local member for Wangaratta and Rutherglen in 1894, and as MP he maintained his socially progressive bearing in his support for women's suffrage and his lobbying for parliamentary and economic change with the Kyabram Reform Movement and the Citizens

Reform League. Bowser had the support of the Victorian Farmers' Union and was known as a 'country liberal'. His talents as a politician eventually saw him elected as premier in 1917, and in 1927 he was knighted for his contribution to Victorian public life.[2]

In the spring of 1902, the issue exercising Bowser's mind was sludge. After nearly a decade of clear water, the Ovens was as polluted as ever. Sludge was back, and it was coming down the river from dredge operators above Myrtleford. Dredging was the new technology reviving the mining industry all over Victoria, instigating another mining boom.

The first bucket dredges used on Victorian rivers were introduced around 1899.[3] At the peak of industry in 1911, there were nearly sixty of them operating around the state. By the end of the 1920s more than 100 had eaten their way along the valleys, and the last ones didn't stop until the 1950s.

John Bowser

As we saw in Chapter 2, bucket dredges were self-contained float-ing factories used for processing vast quantities of low-grade sands and gravels in river valleys. As a dredge advanced it filled in the pond behind it with discarded sludge. Dredges could devour hundreds of acres of rich, fertile valley land. The bucket dredges that appeared in Victoria at the turn of the century could dig to a depth of twenty-five feet (eight metres), bringing deep alluvial gravels to the surface and leaving behind a shingled wasteland – unsightly piles of cob-bles and sand in which nothing would grow. Farmers called them 'desolating dredges'. A government board set up in 1914 labelled it 'industrial vandalism'.[4]

The anti-dredging outcry was practically immediate. Down-stream, communities were upset at the renewed flows of sludge. The Ovens River was 'pea-soupy' and thick with tailings above its junction with the King.[5] Silt and pollution in the Loddon was dam-aging land below Newstead, and floods brought down piles of gravel that choked the river.[6] Farmers and graziers were devastated by the wanton destruction of prime agricultural land on the river flats. The consequences of dredging were summarised by a government report in 1914, which, agreeing with most contemporary agricultural experts, observed that 'many patches of rich soil have been upturned and ruined' for cultivation. These 'patches' amounted to over 4000 acres of land between 1905 and 1913 – some of which was ground previously dug over by miners in the 1850s and 1860s, but much of it hitherto untouched.[7] Dredge operators claimed to be able to restore the land. Others remained unconvinced: 'Rich lands can by dredg-ing be ruined in perpetuity'; '[no] forms of restoration work so far devised ... are of practical and permanent avail'.[8]

Landowners all over the state rallied in opposition to dredg-ing at its commencement in 1899. They sent a large delegation from the upper Murray, Ovens, Kiewa, Loddon, Goulburn and Mitchell valleys to Melbourne with petitions to the premier and the minister for mines. This was only the first of many such delegations.[9] Landowners in Wangaratta discussed legal action to prevent dredging on the King River, and in 1901 the Kiewa Valley Anti-Dredging League began writing to local councils. In June 1902, the Corowa Progress Association convened town meetings in Albury about sludge in the Murray and began lobbying local politicians.

The anti-dredging movement mobilised a broad coalition of stakeholders. One 1907 delegation was a who's who of northeast Victoria and even some interstaters. It included MPs from Wangaratta and Benalla, and representatives from the Shires of Rutherglen and North Ovens, the Rutherglen Waterworks Trust, and the Victorian Piscatorial Council; interstate delegates came from the NSW Fisheries Board, the Corowa Anglers Club, the MP for Corowa, and the Albury Council.[10] These groups found common cause with farming communities who feared for their land and livelihood: in the Goulburn valley, for example, graziers at Alexandra, farmers at Seymour and irrigators at Nagambie were up in arms about sludge flowing from mines in the hills above what is now Lake Eildon; Gippsland farmers were angry about sludge from sluicing and dredging along the Mitchell River; the Bairnsdale Waterworks Trust was particularly vocal; landowners along the Loddon from Yapeen down to Eddington protested the waste coming from Castlemaine. People in these communities closely followed news reports of events in other valleys, held public meetings, corresponded with each other via shire councils, and sent delegations to ministers.[11] After half a century of sludge damage, a substantial movement of people across Victoria and beyond was uniting to take effective action.

Bowser was on the frontline of these protests. As a landowner, his experience of sludge damage was firsthand. As the member for Wangaratta, he received deputations from constituents and attended meetings and rallies on their behalf. He was also well placed to get something done. As a former newspaper owner, he knew how to get stories into the press; as past president of the Country Press Association he had a network of journalist colleagues across the state.[12] And now, as a politician, he decided to force the issue by bringing a private member's bill before parliament in October 1902. The timing was opportune, with south-eastern Australia entering its seventh year of drought, dust storms and water shortages. The bill 'to prevent the pollution of rivers and other water courses with sludge' didn't get beyond the first reading, but it was an important first step, and enough to get the attention of Ewen Cameron, the minister for mines and water supply. Cameron was already planning to revise the *Mines Act*, last visited in 1897 and in need of updating. Between community lobbying and Bowser's anti-sludge pollution bill, there was little doubt about the flow of public sentiment on the sludge question.

Cameron spent much of 1903 touring Victoria to meet with those in the mining industry and seek public input on revisions to the *Mines Act*.[13] A statewide conference of industry bodies was held in autumn. It included the Mining Managers' Association, the Amalgamated Miners' Association, the Legal Managers' Association, the Mine-owners' Association, the Mining Boards and the Chamber of Mines, and several politicians.[14] By the time Cameron introduced the new bill to parliament in October he was well informed about the issues. He contrived that bill include a number of resolutions passed at the industry conference, and so the proposed legislation was generally welcomed by miners. To their surprise and consternation, however, Cameron also included a section – the controversial clause 49 – restricting the pollution of streams by sludge. Bowser's earlier, unsuccessful bill and the widespread public complaint had had an impact.

The Precious Metal, *by J.A. Turner, 1894.*

*Map of Victorian mining districts, Victoria Department
of Crown Lands and Survey, 1866.*

VICTORIAN WATER SUPPLY
PLAN SHEWING
AURIFEROUS GULLIES
IN THE
CASTLEMAINE & FRYERSTOWN MINING
DIVISIONS
COMMANDED BY THE
COLIBAN SCHEME.

Scale - 40chs to One Inch

Clockwise from top left: Victorian water supply in the Castlemaine and Fryerstown mining divisions commanded by the Coliban Scheme, 1871.

Garfield waterwheel stone foundations, 2013.

Loddon Company siphon pipe near Fryerstown, 2013.

Ballarat sludge drain, 2012.

Creswick water race.

SLUDGE

1852

STANLEY, OVENS.

Top: Stanley, Ovens, *by Edward Hulme, c.1859.*

Bottom: Golf Hill Estate, Leigh River, 1909.

The anti-sludge provision was a single clause in the new bill, but the backlash from mining interests telegraphed its significance. Members from mining electorates, the industry and the press erupted in a cacophony of protest.[15] 'It will be a bad day indeed for all connected with mining', said Albert Harris, the member for Gippsland's mountain mines. Frederick Hickford from East Bourke agreed: 'No one wants to see any stream injured by sludge or discoloured water but at the same time we have to be extremely careful to see that mining operations are not interfered with'.[16] Dredge operators on the upper Ovens were outraged. The Bright correspondent to the *Ovens and Murray Advertiser* claimed that

> no new matter is deposited in the water course [by dredging]. The gravel, sand, soil, etc. are only transferred from one position to another ... The worst that has been proved against the Ovens River sludge is that a few holes in which a stray crayfish and an occasional Murray Cod might be captured have been silted up ... Is the great industry that has done so much for the country to be penalised on account of providing an occasional afternoon's fishing during the season?[17]

Metropolitan readers were so stirred by the issue that in November 1903 *The Argus* sent its own reporter to Wangaratta.[18] He spoke to several local people and visited some of the Tarrawingee farms to investigate the sludge for himself. He was shown a fence that stood above two prior fences buried in the sludge beneath it and 'a raised flat several hundred acres in extent ... created above an old series of lagoons ... [with] the level of the ground raised from six ft to ten ft.' The reporter investigated industry claims that sludge was actually good for the soil and found that 'On this point both Mr Chandler and Mr D Diffey, another prominent settler, are emphatic.'

'It is not the river gravel or the light capping of loam off the river flats that does the damage,' said Mr Diffey, 'but the red soil. That is a different material altogether from the sediment brought down a stream by an ordinary fresh. The deposit from cultivated ground is washed down to and fertilises the roots of the herbage. But the red soil lies like cement over the land, and months elapse before the ground recovers itself.'[19]

Dredge operators blamed the sludge on sluicers like John Pund in the Beechworth hills, but the *Argus* reporter saw for himself that the Ovens was polluted above the Tarrawingee sludge channel, well before Pund's waste even reached it.

In parliament, Alfred Bailes and John Fletcher immediately requested a week's adjournment to confer more widely.[20] This would give members the chance to consult with their local electorates. Fletcher wanted to talk to his constituents in the mines around Beechworth. Bailes, from Bendigo, sought to give industry representatives the opportunity to air their views. He also wanted to gather together those in the House who represented mining districts so that they could settle on a collective strategy. The delay gave miners time to rally: those from Bendigo and the Ovens sent petitions, while those in Castlemaine lost no time relaying their fears to the local member.[21]

Discussion was heated when debate resumed in the legislature on 21 October. Most speakers were from mining districts, caucused together under Bailes' leadership. A handful of mining members spoke to the bill but had nothing to say about sludge. These were mainly representatives from drier areas in the west of the state, such as Hugh Menzies from Stawell and George Mitchell from Talbot and Avoca. David Sterry, representing the men who worked deep underground in Bendigo's steaming quartz mines, was more concerned with health and safety issues than with water pollution.

Most of the thirty-one members who spoke, however, had quite a bit to say about the sludge clause. The first speaker, Bailes, began by rousing his colleagues: 'I would like every member representing a mining district to address himself fully to these points, so as to show honorable members that it is not a one-horse affair, and that there is a general unanimity of opinion amongst mining members against that proposal.' He was firm that nothing should be allowed to interfere with mining and any attempt to do so was simply mischievous:

> I am not selfish enough to contend that because some gold may be obtained from a certain mine other people's rights and means of living are to be interfered with in order that that gold may be won ... [B]ut I am rather inclined to think that the feeling of opposition to the mining industry which is shown by people who are on the land, who do not like the idea of the gold-seeker coming near them, prompts them to exaggerate in no slight measure the harm they contend is done to them by the so-called pollution of streams.[22]

A chorus of agreement followed these sentiments.

Several members pointed to the depressed state of mining, arguing that this was not the time to impose further restrictions. Charles Andrews from Geelong argued that 'If we pass such a sweeping provision as the Bill contains to prevent dredging in our streams, we might do a great deal to fetter the mining industry, and to lessen the opportunity for employment to be derived therefrom'.[23] Dredging, specifically targeted by the bill, was seen as the industry's saviour. Sir Alexander Peacock, Creswick native, leader of the Opposition and one-time manager of several of the Berry deep lead mines, had personally seen dredging in New Zealand while on his honeymoon. He was an enthusiastic advocate and told the House that 'The immense

development of dredging for gold in New Zealand is simply marvellous, and what has transpired on the River Ovens has been a perfect
revelation.' Charles Shoppee read a letter from the Ballarat Mining
Board declaring that the bill would 'stop all dredging and nearly
every mine in the Ballarat district', and his fellow Ballarat member,
Joseph Kirton, agreed: 'A more inopportune time could not have
been selected for the introduction of such a clause as this'.[24]

It was not only the dredgers who were alarmed. Hay Kirkwood
from the Bendigo electorate of Eaglehawk thought it would have
a severe effect on quartz crushers. Others spoke in defence of the
independent miner, 'the small man', 'the miner's right man', 'the tub-
and-cradle man', a nostalgic figure evoking the early gold days in
which so many of those present had participated as younger men.
The romantic mythology of the gold rush was already being deployed
in the service of concealing the realities of the mining industry.
George Prendergast, though representing the metropolitan electorate of North Melbourne, had once been a miner himself; he spoke
about clause 49 from the heart:

> It is directed at the small man, to the fellow who is making a liv-
> ing with the cradle and tub, and who could not by any means, if
> he were to work all his natural life, fill up some of the holes in
> the creeks … we should not have legislation which will strike at
> the poor men who are fossicking for a living, and who are doing
> little damage.[25]

Bailes agreed. 'I am sure it never occurred to [the minister] that it
could be applied to the sluicer and the tub-and-cradle man, but it
can', he said.[26]

Several speakers reminded the House of another group of interested players, less romantic but still important: investors. Charles

Hamilton of Windermere stated that, 'It is plainly palpable that it is impossible to ignore the interests of the investing public in any legislation affecting the mining industry'.[27] Fellow investors, including the members for Horsham, Ararat and Stawell, heartily agreed.

Bowser led the case for the rivers. Undeterred by the failure of his private member's bill twelve months earlier, the member for Wangaratta now deployed his considerable skills in the anti-sludge cause. He could hardly contain his passion:

> This Bill has excited the widest and most earnest attention throughout the State ... [It] provides, for the first time in the history, I believe, of this House, for the control, if it should be carried, of the sludge pollution of our rivers in the mining districts ... I hold that an obligation is laid upon this House to see that the injury now being done is not continued, and is not to be written permanently on the face of this State.[28]

He was supported by his colleague from Echuca, John Morrisey, who knew the hardship of farmers struggling through the Federation Drought. Morrisey was equally passionate and eloquent:

> We cannot do too much in connexion with maintaining, in their entirety, the copiousness and the capacity of the streams which we have in Victoria to-day. The last few years have given us only too practical and sad an experience of the necessity of utilizing to the fullest extent the limited water supply we have. I do not suppose on the face of the whole earth there is a worse river system than in Australia, and knowing the value of water and its necessity, we cannot be too guarded in any action we take to protect the streams.[29]

Bowser was worried about farmland at Tarrawingee, but he was also concerned about municipal water supply for Wangaratta. James Graves, too, feared that sludge from mining in the hills of his Delatite electorate was polluting the water supply of Violet Town and Euroa. Joseph Brown from Shepparton felt there was injustice in the discrepancy between the protection of Melbourne's water supply and conditions in country areas. As he pointed out, Melbourne people drew their water from the remote and unsullied tributaries of the Yarra and Plenty rivers, whose catchments had been closed to human occupation for years. Nothing similar existed in country districts and Brown didn't think that was fair at all.[30]

Beneath the general alarm at the potential threat to industry, there were two issues at the core of these debates. The first was the need to protect the rivers, which was agreed in principle even by many of the pro-mining speakers. Peacock conceded that 'reasonable restrictions should be imposed on those engaged in the industry, so as not to destroy any more land, and we know that some of those who have been engaged in the industry in the past have been regardless at times of the interests of other people.[31] Fletcher admitted that in his own district around the Ovens, 'I have seen dredging operations carried on in a careless blundering kind of way that has done a good deal of injury'.[32]

The second, and more intractable, issue was which rivers to protect. Bowser himself conceded that 'We know, unquestionably, that there are many rivers and creeks in our mining districts that are already destroyed and that it is impossible for us to adopt any remedial measure ... Those streams have already been sacrificed, and we may as well give up hoping to restore them.'[33] Others agreed. Certain rivers had been receiving sludge for so long it would be unthinkable to stop. As Fletcher put it, 'Mining has been going on there [at Beechworth] and the water has been polluted since 1852'.

His colleague Thomas Ashworth, representing the Ovens, Peacock and his neighbour Walter Grose from Creswick, and the Castlemaine member Harry Lawson all agreed that something should be done to preserve the rivers – so long as it was not those rivers used by their constituents. As Grose put it, the 'Bill [would] apply not only to streams which today are pure, streams where the water is good, but also to other mining creeks, which have been running sludge for the last twenty or thirty years, the clause was made too wide'.[34] Should the large rivers used for navigation and water supply be protected, which the miners felt would be acceptable, *as well as* the smaller tributary streams, as advocated by those like James Graves from Delatite?[35] Surely this was going too far.

A key difficulty was the bill's terminology. In drafting the clause protecting streams from sludge, Cameron proposed penalties for anyone found to have 'discoloured the water in the creek'.[36] But there was universal agreement among miners that it would be impossible to mine without some disturbance that would cause discolouration. Richard Toutcher from Ararat argued that 'In the case of any river or creek or rivulet, any water course, in fact, there is a little sluicing water, or a little discoloured water, that finds its way into what might be a comparatively useless creek – that is to be considered sufficient to practically annihilate all dredging and all sluicing.'[37] The miners shared this rather outraged view, but so actually did those wanting to protect the rivers. Graves from Delatite, Wallace representing the Loddon farmers, and Morrisey from the irrigation districts around Echuca all agreed that they didn't mind a little discolouration.[38]

The speeches make clear that there was broad agreement on two fundamental points: rivers and farming interests should be protected from mining pollution, and mining was too valuable to stop. The real stumbling blocks were the same ones that had dogged the legislation in 1881 and in 1885. Exactly *which* rivers should be protected? And just what was pollution anyway?

After weeks of debate all through the spring of 1903, Cameron could see the writing on the wall. Christmas was approaching and with it the end of the parliamentary session; if he wanted to get the bill passed, including all the other clauses that had already been agreed to, a new strategy was needed, so he withdrew the sludge clause. Alas, it was already too late for Cameron's proposed legislation: the bill lapsed as the members left for their summer break.

During the recess, the premier resigned and cabinet was reshuffled. When the bill was reintroduced in July 1904 it was sponsored by the new minister for mines, Donald McLeod from Daylesford. Most of the provisions closely followed those in the abandoned 1903 bill and passed without incident. But McLeod, while representing the once-great mining centre of Daylesford, had not forgotten the pollution problem. Mining in his district had long been in decline and the region now depended on agriculture. In the previous years' debates McLeod had spoken passionately about the damage done by sludge: 'I think it is almost impossible for anyone to form a proper estimate of the enormous damage that is being done in this country through the sludge nuisance.'[39] Not to be deterred now, his new bill reintroduced the sludge control clause abandoned by Cameron. 'It is possible to pay even too high a price for gold', McLeod declared. 'It has, therefore, been thought desirable to bring the matter within legal bounds, and enable dredging or hydraulic sluicing to proceed where it will do no damage, and to control it where it is likely to result in damage to the lands along the streams or rivers.'[40]

Debate resumed where it had left off, with members lining up as before to support mining on the one hand or agriculture on the other. They rehearsed the old arguments: mining was in a perilous state; regulation would be an impediment; the sludge clause would make miners guilty of a breach of the law every day of the week. Donald Fletcher had retired, but his Beechworth electorate was now represented by Alfred Billson, who was just as vigorous in his insistence that if rivers were already polluted, it should be permitted that they remain so.[41]

This time, however, Bowser and his allies were even more determined to protect water. They made an explicit connection between the state of the rivers and the future of the irrigation agriculture industry. Bowser summed up their case thus:

> Victoria was already scantily supplied with water. In the arid districts especially there were large towns growing up which were dependent upon the water supply of these rivers for domestic purposes, and there were irrigation works throughout the State in which millions of money had been invested ... It was necessary, if the reservoirs were not to be silted up, and closer settlement and production restricted, for Parliament to afford some protection to the rivers in the State which were all too scanty now for the purposes for which they would be needed in the future.[42]

Speaker after speaker emphasised the amount of public money already spent on water supply infrastructure for towns and agriculture, and the danger of reservoirs filling with sludge. Morrisey from Echuca told parliament that 'Considering the amount of money involved in connexion with water distribution, he felt that in that respect alone action must necessarily be taken by the Minister of Water Supply'.[43] Of the Mitchell River, James Cameron from Gippsland warned that 'If honorable members were not careful in dealing with this subject, the whole of that land would be destroyed, and in doing that they would be destroying something which would otherwise go on producing food for the people for ever'.[44] Toutcher from Stawell was compelled to agree: 'The whole of the agricultural development of the State depended on the prevention of river pollution'.[45]

After modest beginnings in the 1880s, irrigation was steadily growing in Victoria. By the turn of the century, major works included the Laanecoorie Weir on the Loddon, the Goulburn Weir at Nagambie and construction of the Waranga Basin. Local irrigation trusts also developed works on natural streams to carry water, and thousands of farm families in the irrigation districts now depended on the water diverted through these systems. The early weirs were low, raising the river level so water could be distributed down channels, but they were high enough to catch any sludge that flowed in from rivers. If the reservoirs filled with waste, then their capacity to store water was reduced and the money spent on their construction wasted. Water supply was also taking on new national significance after Federation in 1901. Under the Australian Constitution the Commonwealth was responsible for river navigation, but otherwise the rights of the states 'to the reasonable use of the waters of rivers for conservation or irrigation' could not be abridged.[46] The states had promptly begun wrangling over how much water they should get, especially from the Murray and its tributaries – an argument that continues to this day.

The men debating water in the Victorian parliament in the winter of 1904 were acutely aware of the effects of the Federation Drought, one of the worst droughts ever experienced in southeastern Australia.[47] Rainfall had begun to decline in 1895 and by 1902 the country was in crisis. Special trains carried water to stricken families in north-western Victoria and politicians like Bowser participated in local fundraisers to assist. Crop production fell to fifteen per cent of average yields, sheep and cattle died in their thousands in barren paddocks, and water in the Goulburn River fell to twelve per cent of its usual level. Strong winds in November 1902 stirred up an immense dust storm that reduced visibility to 100 metres in Melbourne and dumped thirty centimetres of sand on the railway line at Swan Hill. Conditions didn't begin to ease until good rains fell in 1903. The supply of secure water for farmers was on everyone's mind.

Donald McLeod was deeply conscious of both the cost of supplying water and the damage to reservoirs. He admitted to one group of anti-sludge campaigners that the state would soon have spent £8 million on water. He told them that 'The future of the country depends on irrigation and water supply'.[48] Not mining, not anymore. As a Daylesford man, McLeod was also aware of what was happening to the Laanecoorie Weir on the Loddon River. A decade after its construction it already contained three metres of silt in its basin, the result of mining sludge, and this significantly diminished the amount of water it could hold.[49] McLeod told the House: 'Parliament had spent a lot of money in providing a water supply on some of those rivers, and common sense would show that it could not allow those rivers to be spoiled by the deposit of silt.'[50]

The irrigation argument was a powerful point. But the pro-river faction had another new card to play. They had been busy since the last bill lapsed investigating that other great Pacific Rim mining region, California, and had learned some things from the American experience. This time the anti-sludge provisions in the bill centred on the creation of a Sludge Abatement Board, modelled on the California Debris Commission, which was established following the cessation of hydraulic mining in 1884 and its cautious, regulated re-emergence under the federal *Caminetti Act* (*California Debris Commission Act*) 1893.[51]

Bowser had been planning the Sludge Abatement Board for some time, hoping to get it included in the 1903 bill. At the start of the debate the previous year he gave fellow parliamentarians a history of the sludge problems in California and described the California Debris Commission in detail. He summed up by saying:

> I propose, when the Bill is in committee, submitting a series of amendments, with the object of following the Californian system of providing for the creation of a Sludge Abatement Board. The board may be composed of the warden for the district, the Inspector General of Public Works, and the Chief Engineer of Water Supply or the Secretary of Mines, who shall be appointed by the Governor in Council. The intention of the amendment is that the board shall see all complaints made from time to time, receive and hear them, make an inspection of the locality, and determine upon sites for settling-dams where they are necessary. The board will determine whether the dams are necessary, and then select sites, receiving plans and specifications, which have to be approved by them. They will also have to see that the works are efficiently carried out. The amendment also provides for certain penalties, and that the board may exempt from the operation of the Act certain rivers and water-courses which have already been injured by sludge pollution, and which it is hopeless to attempt to reclaim.[52]

As it happened, parliament was adjourned before Bowser had the chance to introduce his amendments. But he made sure they got into the new bill.

The provision for a Sludge Abatement Board in the 1904 bill was an inspired stroke. It delegated responsibility for administration of sludge control to the Board so that the bill itself could be framed at a level general enough to apply to the whole state. Specific conditions in each region could be assessed and regulated by the Board, thus providing the flexibility to account for the diverse circumstances of local industry and topography. Members could vote in favour of clean water in the expectation that the rivers they knew could be exempted under the Board's jurisdiction. The quantity of sediment allowed could be defined separately to the quantity of poisonous matter so that the deep quartz mines would not be affected. At long last a way had been found to resolve the impasse between the economic benefits of mining and the environmental cost of sludge. It still required weeks of debate, but in mid-December 1904 the bill finally passed into law as the updated *Mines Act*. Victoria finally had legislation that would stop the sludge and protect the rivers.

When the Sludge Abatement Board was formally commissioned in 1905 it looked remarkably like what Bowser had envisioned. It consisted of three appointed members representing water supply, land and mining. It would establish regulations to prevent debris from entering waterways, review lease applications to ensure appropriate control measures were in place and investigate complaints. There would be formal definitions of sludge based on the quantity of suspended sediment and on the amount of 'poisonous matter' it contained, and the Board had the power to refuse lease applications if the threat of damage was too great. Similarly, it had the power to exempt certain rivers from the provisions if they were deemed already polluted.[53]

Bowser's predecessor and fellow river defender Henry Parfitt didn't live to see the vision of the old 1886 Sludge Inquiry Board realised. He'd died a decade earlier, not long after leaving parliament. Parfitt's Tarrawingee neighbour Hopton Nolan saw it though. Old Hoppie was seventy-seven by then, but still going strong. He continued to run the Plough Hotel until a year before his death in 1910, and was certain to have followed these proceedings with interest. It's nice to imagine that Bowser stopped in at the Plough for a drink to celebrate their victory.

But Bowser and Nolan and their allies did not celebrate for long. It was soon apparent that there was more work to be done. They didn't know it then, but it would be many years yet before rivers would run clear. For one thing, it took some time before the Sludge Abatement Board was fully operational. It also took time to convince the mining industry that the Board was serious.

Before the Sludge Abatement Board was even set up, the government got busy reassuring the industry that the new system would not spell the end of mining. Minister for Mines Donald McLeod hurried off to the goldfields as soon as the bill was passed to talk to miners. Early in the new year he was in Beechworth reassuring an audience of sluicers and dredgers that 'mere discolouration of the water' would not be a problem and that 'the Government would be guided by the special circumstances surrounding each particular case'. A reporter for the *Ovens and Murray Advertiser* was not convinced: 'The unfortunate fact is that the power of the Minister is restricted in this matter. The sludge clauses are under the jurisdiction of the Sludge Board ... which may conceivably ignore any suggestions or recommendations he has to offer. Of what value in that case are all his assurances?'[54]

When the anticipated crisis came it wasn't on the Ovens; it was on the Loddon in McLeod's own Daylesford electorate. Grazier Alexander Clarke owned land at Joyces Creek just downstream from Newstead. He bought it in 1901, when the river had been clear. Less than eighteen months later the water was so 'thick and muddy' his cattle could no longer drink it. Three feet of silt settled in the river-bed so that 'cattle wading through it came up to their bellies' and the dried mud stuck to their sides like cement.[55] Clarke concluded that the sludge was coming from the Loddon Gold Dredging Company above Guildford, and in the autumn of 1905 he filed an injunction in the County Court to stop it.

SLUDGE ABATEMENT BOARD

Sludge Abatement Board letterhead

The case threw the industry into confusion. Thirty witnesses (including then civil engineer John Monash, well before his fame as military leader during the First World War) gave evidence over three weeks of hearings. Proceedings were widely reported in local and metropolitan newspapers. The dredging company argued that a custom had emerged over the years whereby miners were entitled to pour their sludge into creeks and streams. Judge Chomley disagreed and found in favour of the plaintiff, but awarded him just £25 in damages, a fraction of the sum he'd sought.[56] When the judgement went against the miners the *Bendigo Advertiser* led its report with the headline 'Industry Seriously Threatened'.[57] The Castlemaine Dredging Association pledged to cover the legal costs of the companies involved and organised a meeting in Melbourne to rally more support. The meeting attracted mining delegates from Ballarat, Buninyong, Enfield, Maryborough, Avoca, the Ovens, and Bendigo, and was attended by Alfred Billson. Sir Alexander Peacock and five other politicians sent their apologies. All agreed this was a test case and pledged to form an association to protect the industry.[58] Heartened by the support, the Loddon Gold Dredging Company appealed the decision in the Supreme Court and the fight went on.

Alexander Clarke, however, had allies of his own. As a councillor for the Shire of Newstead he was a man of local influence with neighbours who shared his concerns. By September 1905 they too were organising an association 'to take steps to prevent the further pollution of the stream'.[59] The farming community met at Newstead and further down the Loddon at Baringhup and Eddington. By December, landowners along a forty-mile stretch of the river had formed the River Frontages Anti-Pollution Association, soon renamed the Loddon Anti-Sludge Association. The dredging companies requested a meeting with landholders and in January 1906 the farmers toured miners around the district, showing them the silt that filled the Laanecoorie Weir. The dredgers acknowledged that much damage had been done, that it was difficult to water stock, and that they 'should be unanimous in working on a scheme that would prevent further trouble'.[60] The visible damage must have made an impression, because the Castlemaine Dredging Association followed up the visit with a letter to its members, urging them to do everything they could to reduce the sludge. Then, 'Owing to the friendly understanding now existing between the landowners and the Association', it dropped its legal appeal.[61]

Understandings were not friendly enough, though, and only a few months later Clarke returned to court seeking further injunctions. Both sides took their case to the minister for mines in a monster meeting in June 1906.[62] The dredgers were represented by the presidents of the Dredging Association and the Chamber of Mines, along with Harry Lawson, the state member for Castlemaine, and federal MP James McCay, as well as the lord mayor of Melbourne, among many others. Landowners had also drawn support from further afield: in addition to three state politicians, they were joined by representatives from the Shire of Shepparton, the Elmore Water Trust and the Nagambie Water Trust, plus residents from Bet Bet and Tullaroop along Deep Creek and from the Leigh River valley south of Ballarat. The farmers didn't bring their local member because they were represented by McLeod, the minister for mines himself, who made his position clear at the meeting: 'The sludge board said there had been deliberate wrongdoing. The dredging companies did not come to him with clean hands... The future of the country depended on irrigation and water supply.'[63] Farmers had also won the support of the metropolitan press. In April 1906, *The Age* editorialised: 'The Loddon River is a considerable stream, and its permanent value for agricultural purposes is vastly greater than that of all the gold to be found in its bed'.[64]

McLeod was in fact already preparing to return the *Mines Act* to parliament for further amendment that would strengthen the powers of government over dredging. After only a year of operation the Sludge Abatement Board had gathered the ammunition he needed. It had found numerous cases of operators flouting their responsibilities and had started to act against offenders. On the Loddon it found that many companies stack sludge 'only till the thick water overflows into the creeks ... occasionally large quantities of slum escape [because of breaches during floods] ... [and] dredge masters sometimes deliberately discharge sludge into the watercourses to save the trouble and expense of stacking it'.[65] McLeod pointed out that 'at the time we passed the 1904 Act ... it was thought that we were overlegislating', but experience was quickly proving that they had not gone far enough.[66]

One problem was the difference between mining leases and mining claims. The Board was discovering that the 1904 *Mines Act* only applied to dredgers who had taken out leases – and not those working under miners' rights claims, the old system dating back to the post-Eureka reforms of the 1850s. Miners' rights were intended to allow small-scale, independent miners the opportunity to make a living without the obligations imposed on big company leaseholders. But claimholders worked around the system by consolidating claims into large blocks of up to 200 acres.[67] Claims could be registered with minimal oversight, so dredge operators were able to avoid the Sludge Abatement Board's regulations. McLeod complained that as minister for mines he had refused dredging leases because of the risk to the river, with companies then 'coolly taking up land under consolidated miners' rights'.[68] The amendments he had in mind would put an end to tricks like that:

> What we provide in the Bill is that if anyone takes up land for dredging purposes he must take it up under a lease, and if anyone wants to hold the land as a claim he must not work more than 5 acres ... we will not have one set of men holding 120 or 150 acres bound down by strict conditions under a lease, and another party alongside working an area of the same size which they have obtained by consolidating miners' rights, under no obligations at all.[69]

The members for Ovens and Castlemaine immediately protested these 'drastic actions'. Harry Lawson argued that it was entirely unjustified to expect small cooperative companies to go to the expense of having leases surveyed. Alfred Billson thought the Sludge Abatement Board already had far too much power – exercised in his district 'up to the hilt'. He claimed that 'They had issued orders which made it almost impossible for the companies to carry on in some instances.' Lawson and Billson, however, were the only two to speak against the provisions; Billson was quite correct when he opined that 'my voice is that of one crying in the wilderness'.[70] Other members from mining districts recognised by now that most reputable miners were in fact able to meet their obligations and continue operating. They could see that a few rogue operators were making it difficult for others, including 'the old people' who wanted 'their rights as fossickers' preserved, as well as larger dredging companies.[71] As the Sludge Abatement Board was also finding, communities were objecting to lease applications because 'the residents have read that injury has been caused in some other locality … Thus the offenders against the law in any one place are … prejudicing the prospects of mining throughout the State.'[72] Legislators accepted the tightened regulations, and the new bill was passed as the *Mines Act* 1907.

Even with increased powers, the Sludge Abatement Board faced a long struggle to get the state's miners to take responsibility for sludge (see figures on page 202). After two years of work the Board still complained that 'mining managers professed that they had not been apprised of the steps ordered to be taken, and were in ignorance of the provisions of the law'.[73] High on the list for its first year, 1905, was the Loddon Valley where it found 'deliberate wrong-doing' and the problems that led to Alexander Clarke's complaints.[74] Bendigo and the Mitchell and Mitta Mitta rivers were also priority areas. Visits to the Yarrowee/Leigh south of Ballarat, the Ovens and the Murray followed in 1906, and in 1907 they tackled the backlog of lease applications on the Goulburn. The Board started prosecutions against some offenders but more often they issued warning notices accompanied by instructions about what needed to be done to comply with the new Act. During its first three years the Board issued 144 orders around the state.[75]

Quartz crushers at Ballarat also proved recalcitrant.[76] In 1906 the Sludge Abatement Board discovered that not only did they make little effort to stack their tailings but, they were also abetted by the local council, who didn't object 'so long as it is taken beyond their municipal boundary'. The Board found 800 acres of prime floodplain down the river at Shelford covered by up to five feet of fresh sand and sludge, nearly forty years after George Russell and the Berthons had first complained. Eight orders were issued to Ballarat companies in 1906 and fifteen in 1907. The following year, there were twenty-two.

Bendigo sludge lake and mining operation, c.1925–c.1940

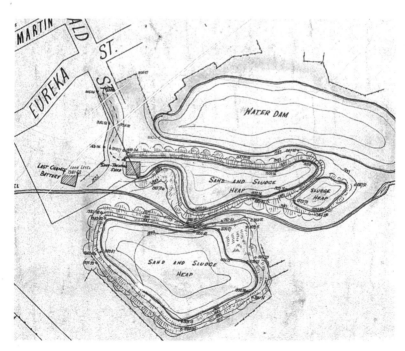

Map of Ballarat East sludge dam, 1905

A test case was dismissed on a technicality, after which some of the miners were 'emboldened by the failure of the Ballarat prosecutions' and relaxed their efforts.[77] The struggle continued the following year when companies again used technicalities to successfully challenge prosecution efforts. The new procedures gave companies another loophole: the Board now had to give a week's notice that it would be serving orders, and the companies used the intervening period to flush out their waste.[78] This happened in March 1910: the Board gave notice that they would be serving orders, but before the orders were officially served a big storm occurred and a company flushed thousands of tons of additional sand into the creek. Nearby at Creswick, several dredging companies also took advantage of the situation to surreptitiously rid themselves of large quantities of waste.[79] Despite these frustrations the Board persisted and at last in 1911 was able to report that 'the Yarrowee River is now receiving very little sand or sludge from the mines'.[80]

Gradually, the Sludge Abatement Board worked its way around Victoria. It started taking samples of water quality too, publishing thousands of measurements from mining-affected rivers across the state. By 1914 the Board had compiled data on the tributaries of the Loddon that showed a significant decrease in pollution from Castlemaine down to Newstead.[81] Community vigilance and complaints continued, however; and, after Alexander Clarke died in 1909, the farmers' cause along the Loddon was taken up by S.J. Oxley from Strangways. Oxley kept a close eye on the miners and did not hesitate to let the Board know when things were going wrong. In north-east Victoria, farmers in the Ovens Valley formed the Ovens River Frontage Anti-Sludge Pollution Association, a community group that remained active into the 1930s.[82]

By 1913 there were still enough concerns about the sludge nuisance for the government to launch a Board of Inquiry into Complaints of

Injury by Dredging and Sluicing. Members of the previous inquiry, in 1886, had all been appointed on the basis of extensive mining experience, but membership of the 1913 Board of Inquiry was much broader. In addition to Mr Merrin from the Mines Department and Mr Gray from the Chamber of Mines, it included representatives from the Department of Agriculture, the Lands Department, the Australian Natives Association, Waterworks Trusts, and the State Rivers and Water Supply Commission. The Board of Inquiry's activities were of intense interest to the public and were widely reported in newspapers across Victoria. Melbourne papers published itineraries and transcripts of evidence given at public hearings. Those in *The Age* were regularly reported under the large-cap headline 'Desolating Dredges'.[83]

It was obvious the rivers were still in a bad way. The Board of Inquiry wrote, 'That river and creek beds have silted up largely ... as results of gold seeking operations goes without saying.' They concluded, however, that most of the damage had already been done by 'the irresponsible methods of the diggers of the earlier days'. After visiting twelve mining districts in Victoria and three in New South Wales, inspecting forty-five bucket dredges and hearing from 112 witnesses, the Board of Inquiry concluded that 'under the present practice of stacking tailings as provided for by the new conditions, the possibilities of escape of earth or *débris* in large quantities ... should be comparatively trifling'.[84]

The Sludge Abatement Board was finally having a potent effect.[85] At Ballarat, where as recently as 1909 companies were deliberately flouting the regulations, 'this evil has practically ceased ... there is now practically no discharge'. Even the notorious hill-sluicers were improving: 'efforts at the conservation of tailings ... and the reduction of material in escaping sluice water are now being insisted on by the Sludge Abatement Board with at least some measure of success'.

There were still some offenders. At Eurobin in the Ovens Valley, the Confidence Dredge was re-soiling 'in a manner so crude as to give little promise of restorative results'. John and Percy Pund's operation at Three Mile Creek was offered as an example of ground sluicing that had resulted in 'the filling up of creeks and low-lying alluvial flats, certainly not to their improvement'.

Instances of serious pollution, however, were becoming less and less frequent. The Board of Inquiry was far more concerned about prospects for future mining and the potential damage to valuable agricultural land. 'It is in its permanent injury of valuable agricultural land', they wrote, '[that] this branch of the mining industry demands restriction and control.'[86] They used the Sludge Abatement Board's evidence to calculate that over 4000 acres of land had already been dredged and applications were in hand to dredge nearly as much again. Inquiry members vehemently opposed this destruction, proposing a series of recommendations to prevent further damage.

These recommendations were never considered by parliament, but in the end the rivers were saved from dredging anyway. The report was submitted in January 1914 and extensively reported in the metropolitan and regional press.[87] Newspapers kept up the pressure over the following months. Cabinet endorsed the recommendations and forwarded them to James Drysdale Brown, the minister for mines. Beechworth member Alfred Billson, who had served as mining minister during 1913, was livid and made his views known in parliament, where he was rebutted by John Bowser. Billson was on his own, however, and the Board of Inquiry's report achieved its authors' aims. The Sludge Abatement Board was strengthened, and its role in restraining the mining industry was endorsed.

Then, in August 1914, as the recommendations were still being implemented, war was declared with Germany. Miners enlisted in large numbers, and with a declining workforce the mines started to close. By this time gold yields had been in slow decline for many years. Dredging, the saviour that kept the industry going, had also passed its peak, a high point in 1910–1912 when more than fifty operations recovered almost 60,000 ounces of gold a year. When members of the Board of Inquiry toured the state's mining districts in 1913 they were seeing an industry on the cusp of a rapid decline. Upon the outbreak of war there were still 10,398 men working in Victoria's goldmines; by Armistice Day, 1918, there were only 3547. While in 1913 the Board saw up to thirty dredges operating in the Ovens and Buckland valleys, a decade later there were none.

It was not only the war that saved the rivers, though. The Board of Inquiry of 1913 demonstrated that regulations administered by the Sludge Abatement Board were making a difference. After the war its influence increased. With fewer mining operations to oversee, the Sludge Abatement Board could be more vigilant. At the same time, mining was more likely to be conducted under the stringent requirements of the 1907 Amending Act. Legislation on mining ventilation and explosives had made the industry safer for workers. New bylaws prevented miners from discharging sludge onto roads, gardens, houses or businesses.[88] In particular, the dredging industry was charged with the need to re-soil, and more broadly there emerged an interest in remediating worked ground.

Before the advent of the Sludge Abatement Board no one in mining had given much thought to what happened when work stopped and a mine closed. Machinery was sold where possible, but otherwise the miners just walked away, leaving open shafts and enormous heaps of mullock and tailings to blow away in the wind. By the early twentieth century, however, things had shifted. Dredge operators were expected to repair the ground by advance stripping and stockpiling topsoil and then replacing it on dredged surfaces. There were revegetation efforts to grow pastures of native and introduced grasses, while experiments with fruit trees on remediated ground were promising.

SLUDGE ABATEMENT BOARD
Sec. ....Mines Act....................
TYPE DRAWING № 8.
To accompany Order №...................on............. ....

Retaining dam, Sludge Abatement Board, 1907

Hydraulic sluicers were subject to more environmental controls as well. By 1912, John and Percy Pund had constructed a series of settling basins in Three Mile Creek made from brush to retain the sludge from their sluicing works (see figure above). They constructed 400 metres of embankment from two to four metres in height, impounding an area of almost two hectares.[89] When Pund's leases were acquired by G.S.G. Amalgamated in 1919, the company continued with the use of sludge settling basins in the creek until the close of operations in 1948.[90] These basins, long since filled to the brim with sludge, are well preserved as large grassy terraces in the valley of Three Mile Creek today.

When bucket dredging revived from the 1930s onwards the companies were subject to even stricter operating conditions. A smaller number of larger plants at Newstead, El Dorado and Harrietville were forced to comply as authorities sought to harmonise the interests of both miners and landholders. Greater attention was to be placed on stripping and redepositing overburden, controlling effluent in settling ponds, levelling dredged areas with bulldozers and re-sowing with pasture grasses.

John Bowser thought that saving the rivers was one of the most important things that he had done, and he was convinced that future generations would agree. He told parliament in 1914 that 'I have no hesitation in saying that that work will be looked back to by young Victorians of the future as the most valuable done by any Board or Commission that we have appointed during recent years'.[91] He was right about the significance of his achievements, but he was wrong about how the struggle against sludge would be remembered. As the rivers gradually cleared, the entire catastrophe was almost entirely forgotten. How is it possible that a society can forget an environmental disaster that engulfed an entire state, taking generations to resolve? And does such forgetting matter?

7

AFTERMATH

YOUR SMARTPHONE IS PROBABLY CLOSE BY AS YOU READ this book. It's the mine of the future. It's full of precious metals, rare earths and minerals. Among its constituent parts are gold, silver, copper, aluminium, nickel, selenium, cadmium, antimony, cobalt, lead, lithium, palladium, tantalum and zinc. All these minerals have been extracted from holes dug somewhere in the earth's surface. One estimate suggests that 1.6 million tonnes of rock were mined to produce just the gold in a single model of mobile phones, the world's forty-five million iPhone 5s.[1] Most rare earth minerals come from China, although Australia is growing as a producer, and prospecting continues in Alaska and Canada. Some components are potentially conflict minerals, sourced from war-ravaged countries like the Congo, and sold to perpetuate the fighting.

Mining and mineral processing always leave some kind of environmental legacy. This might be abrasive dust containing microcontaminants, like that which blows through the streets and gardens of Mount Isa and Broken Hill; toxic water like that which sits behind dam walls at the gold mines near Orange in New South Wales; or the acid mine drainage that seeps from the old Mount Lyell copper mine on Tasmania's west coast. It might be the holes themselves, like the Super Pit at Kalgoorlie, a massive void in the earth 3.5 kilometres long, 1.5 kilometres wide and 600 metres deep; or the mountains of waste rock piled up from these holes' excavation. Radioactive waste from rare earth mining can have significant public health and environmental effects. When protestors lobby against mining companies their targets tend to be coal producers. Coal mines are clear targets in an era of climate change and renewable energy, but most mining is driven by other markets that respond to our choices as consumers. Our modern love affair with technology comes at a profound environmental cost.

Today when Australians think about mining and its environmental impact they tend to focus on the big mining states of Western Australia and Queensland. They think about uranium mining in Kakadu, the effect of coal mines on the Great Barrier Reef, and open-cut mines working their way through the picturesque Hunter Valley in New South Wales. People rarely associate Victoria with mining at all, though they might remember the Hazelwood fires of 2014, when a coal mine in the Latrobe Valley burned for forty-five days, leaving local residents and fire fighters alike with lingering health effects.[2] People remember that this state had a gold *rush*, but Victoria's gold mining *industry* has been all but forgotten. We've forgotten the big companies and the huge profits delivered to investors in nineteenth-century Melbourne and abroad. We've forgotten the political and economic clout wielded by industry lobbyists, and the thousands of lives cut short by appalling working conditions. And we've all but erased the environmental disaster that characterised mining districts for more than half a century. Victoria may no longer be one of Australia's big mining states, but it certainly once was, and we are still paying the environmental cost of the days when Victoria was the jewel in Australia's mining crown.

Our selective un/remembering of this past raises important questions: why have we forgotten this history, and what has happened since mining ended to allow us to forget so completely? In this final chapter, we offer some answers to those questions, and we ponder some of the lessons that the sludge problem offers for modern issues of water management and industrial waste. First, though, we must pause and consider another: what is the evidence that the mining industry *has* indeed been forgotten? Is this true, or merely an assertion that suits our own agenda as environmental historians?

One clear sign that Victoria's mining industry history has been forgotten is the way the state handles former mining sites. These are places where the resource has been exhausted or is no longer profitable to recover, and where mining is thus no longer active and not expected to resume. When no owners can be identified, such places are described as 'derelict' or 'abandoned'. They are 'legacy' mines and we collectively inherit the responsibility of managing them. They can be found all over Australia and they are a big, ongoing problem.

Today, the responsible – and often legally mandated – course of action when mining finishes is to decommission the mine. Machinery is removed, shafts and open cuts are filled in, the surface is revegetated, and contaminants are safely secured. The development lease that allowed the mine to operate is not fully relinquished until the mine has been satisfactorily rehabilitated.[3] Resource companies applying for such leases today must produce 'whole of life' plans for their operations that include details of what will happen to the site when the mine closes. In most states, companies must also deposit a bond to guarantee decommissioning will be carried out. In 2017, the NSW government held over $2.2 billion in rehabilitation bonds from operating mines.[4] This sounds like a lot, but shared among the more than 230 still-active or under-lease mines in the state, it's not so much. The Tasmanian government spends approximately $160,000 each year on rehabilitating legacy mines across the state, another significant underestimate of the full cost. A more realistic tally of the cost of mine rehabilitation is the budget allocated for just one abandoned mine site in Tasmania – Savage River in the state's north. The new leaseholder has identified $21 million as the cost of managing acid drainage that has affected thirty kilometres of the river since magnetite mining began in the 1960s.[5]

In many abandoned mines cases no owner can be identified. The NSW government estimates that there are over 400 such derelict mines in the state.[6] South Australia, whose government calls them 'orphaned mines', has an estimated 3600.[7] Where ownership is unknown, responsibility is left with the landowner – typically the government. The result is thousands of abandoned mines where no remediation at all has been done. And the environmental consequences of this are diverse and significant; they include acid drainage that leaches into waterways, heavy metals such as lead and mercury accumulating in soils and moving up the food chain, and the lingering presence of arsenic and cyanide in onsite tailings. These impacts can persist for centuries, if not millennia.

The active management of decommissioned mines is a relatively recent practice. New South Wales has been doing it longer than most, introducing a Derelict Mines Program in 1974. Since then, schemes requiring industry contributions for rehabilitation have been introduced in most states and territories, from Tasmania in 1995 to Western Australia and the Northern Territory in 2013.[8] State and federal governments have introduced abandoned mines policies to manage these issues, including Queensland's Abandoned Mines Land Program, Tasmania's Rehabilitation Trust Fund and South Australia's Extractive Areas Trust Fund.[9] A recent study by researchers at the University of Queensland estimated that there are more than 52,000 abandoned mines around Australia. This is an underestimate, because the available datasets are known to be incomplete and there is no evidence available for the Northern Territory.[10]

Not surprisingly, the results of these schemes are uneven. A 2017 report into abandoned mines in New South Wales carried out by the

Australia Institute found only two sites that either had been or were being remediated.[11] The cost of remediation is high at a stage in the life of a mine when profits are low or non-existent. Unsurprisingly, some operators implement the least expensive solutions rather than apply industry best practice. Many unused mines are held under lease in an indefinite 'care and maintenance' phase in which nothing is done. Some operators evade their responsibilities altogether by declaring bankruptcy and simply walking away. The authors of the Australia Institute report concluded that 'the industry does not have a good record at cleaning up after itself'.[12]

We might imagine that with Victoria's rich mining history there would be many legacy mines here too. This is probably true, but in fact no one really knows – and no one is really even trying to find out. Victoria doesn't have an abandoned mines database, policy or funds allocation. We could find no evidence that it is a member of the Abandoned Mines Working Group, which is a joint initiative of industry peak bodies the Minerals Council of Australia and the Ministerial Council on Mineral and Petroleum Resources. Nor could we find evidence that the state contributed to the group's 2010 policy document, *Strategic Framework for Managing Abandoned Mines in the Minerals Industry*. Victoria manages some old mines as heritage sites and oversees current licences on a case-by-case basis, with bonds and provisions for rehabilitation required as part of the works plan before the licence is granted. Mining is explicitly exempted from having to obtain approvals under the *Environmental Protection Act*.[13]

There are of course some clues about the potential number of legacy mines in Victoria. Beginning in the 1930s, the Mines Department began a program to fill or cap mine shafts on the goldfields, mostly around Ballarat and Bendigo. By 1950 more than 1600 shafts had been either filled with waste rock or capped with concrete. The program was expanded, with heavy machinery used to fill shafts en masse, and when it wound up in the 1970s almost 100,000 abandoned shafts had been infilled.[14]

Information on about 3400 legacy mines is also available in a pair of datasets called VicMine and VicProd, prepared by Geoscience Victoria. Data includes information on the location of historical Victorian gold mines and how much gold they produced. These datasets are useful but they significantly under-represent the full extent of past mining in the state. Most of the sites described, for example, are quartz mines, with only a small number of alluvial operations included. Relatedly, VicMine records mining *leases*, but historically a great deal of mining was done under miners' rights *claims*, which have not been spatially mapped. VicProd includes only a few records from the Ballarat and Ararat divisions, despite their historical importance.

Another important resource is the work of archaeologist David Bannear. During the 1990s, David recorded thousands of mining sites and features around Victoria for their cultural heritage value. His survey was representative rather than exhaustive, but it remains the most detailed record we have of where historical mining occurred in Victoria, and the nature of the surviving evidence. The program recorded 1700 sites, most of which are areas that incorporate multiple shafts and individual mines. Most of David's work was incorporated into the Victorian Heritage Database, an online resource maintained by Heritage Victoria. The database includes claims as well as leases, so it records mines missing from the Geoscience Victoria data. It doesn't include more recent mining activity, or areas that were once mined (and may still have legacy issues) but have no surviving archaeological evidence, like all those shafts filled by the Mines Department between the 1930s and the 1970s. So, evidence for legacy mines in Victoria is patchy and often hard to find, yet it is clear that the remains of many thousands of operations, features and claim areas survive across the state.

The absence of an abandoned mines policy in Victoria has considerable implications for land management and government policy. For one, it means the impact of mining is not routinely considered in baseline scientific studies of water quality, pollution or biodiversity. Areas of potential contamination are thus only patchily identified in local government planning processes and zoning regulations, and catchment managers and landcare groups are missing some of the surviving stocks of seed needed for revegetation programs. And it means that the implications of legacy mining for recreational fishing, food production and public health are not routinely considered and managed either.

Such absences are the shape of what we have forgotten about Victoria's mining industry. A telling example is the inclusion of the Ovens River under the 1992 *Heritage Rivers Act*, legislation that marks out twenty river systems for special protection based on their biodiversity and value as places of recreation and cultural heritage. The Ovens is particularly significant because it's one of the few major systems in the southern Murray–Darling Basin that doesn't have a large dam regulating its flow. The river is indeed precious and beautiful, but it is not in its pre-European condition. From our research we know that the Ovens, together with the Loddon, were among the most heavily mined river catchments in Victoria. The Ovens has been straightened, diverted, filled with sludge and dredged along many miles of its length. Knowledge of this activity need not diminish the river's value, but it does change how we appreciate it. Knowing what we know of the Ovens' mining history, we should value it even more because it is in fact a laboratory for understanding the long-term resilience of river systems affected by mining. But, crucially, it can only provide such insight when we identify and acknowledge what has happened there – not when we allow its history to be forgotten. We can ask what it means that this valley, which has been so damaged, is now so highly valued. We can ask what the Ovens tells us about how landscapes readjust, and what the environmental effect of today's mines will be in the future.

Other silences have become apparent to us as we have researched and written this book, and these have their own contemporary parallels and legacies. The ways in which race, class and gender have shaped the story about water supply and pollution in the nineteenth century is one such silence. You may have noticed that the story we have written is dominated almost exclusively by white men: white men clever enough to profit from water; men confident and prominent enough to make their voices heard in protest; and men publicly empowered to devise and implement the rules that others followed. Women, Chinese and Aboriginal people are harder to find, yet they were also part of the story. It was rare for women or Chinese men to own water licences, but some did, and we have tried to include them as much as possible. We have not found any Aboriginal licence holders who profited from the sale of water, and yet First Nations Australians never ceded ownership. Women, the Chinese and First Australians are hard to see in documentary sources, but they bore the cost of the new systems for managing water.

By the 1870s and 1880s, for example, when water licences and the infrastructure to supply the sluices were both in place, at least one-third of those still working in alluvial mining were Chinese. For the most part they didn't hold water licences, but they were the ones using the water and often paying handsomely for it.

Women were also significant water users, but their needs took second place to mining. By the end of the 1850s, Victorian women and their partners were starting families, and those growing – and frequently large – families needed clean and safe drinking water, but mines were supplied years before the towns. The tragic human cost of such delays is evident in the shocking death rates among children. In Ballarat in 1861 there were 153 infant deaths for every 1000 births, and in Bendigo it was worse, with 225 deaths for every 1000. Dysentery and diarrhoea from unclean water killed children and adults alike.[15] The Dja Dja Wurrung, whose country encompasses goldfields including Bendigo, Castlemaine, Daylesford, Clunes and Maryborough, describe rivers as 'the veins of Country[,] places central to Creation stories' and vital sources of food, medicine and tools.[16] Their pain at witnessing the rivers fill with sludge can only be imagined.

The injustice built into the flows of water on the Victorian goldfields continues to be a feature of access to water around the world today. Access to clean drinking water is a fundamental human right recognised by the United Nations, but for the poorest twenty per cent of the world's population in Africa and Asia, obtaining clean water is a significant economic burden, costing up to eleven per cent of household budgets, in addition to the time and labour of the women and children who collect it.[17] More than 1.8 billion people around the world rely on drinking water that is contaminated by human waste, and they suffer the same illnesses and deaths as Bendigo residents did in the 1860s. Even in contemporary Australia access to water is still shaped by gender and class. A recent study of residents in Melbourne and Perth found that water restrictions imposed during times of drought are felt most keenly by socially and economically disadvantaged households, many of which are headed by women. These households are unable to afford water-saving devices, take advantage of government rebates, or enjoy alternative cooling methods such as air conditioning on hot days.[18]

On the brighter side, there were achievements that have left enduring legacies. One of these was the initiation of new understandings, frameworks and industrial systems that shape our relationship with water – often for the better. Victoria's sludge problems occurred because miners worked to access this vital resource. Access to water was an engineering challenge, but also a legal and social one. By the mid-1860s innovations had been devised to resolve these obstacles, and engineers and surveyors who had honed their skills building dams and races were thus poised to contribute to the construction of Victoria's public water infrastructure, especially the Coliban System.

John Henry Reilly, for example, whom we've encountered throughout this book, offered his plans for the Coliban System in 1864, as did Edward Wardle, who had constructed water races at Daylesford. Peter Wright developed an intimate knowledge of water races around Beechworth and worked on the Coliban as a district engineer in the late 1860s. Thomas L. Browne was the district mining surveyor at Castlemaine for decades, and in 1862 proposed the supply of mining and domestic water to the township. Robert Clark worked for many years as a miner at Bendigo, fighting for the eight-hour working day, and in 1880 became Victoria's first commissioner of water supply. Previously, the Water Supply Department had been a branch of the powerful Mines Department, and the two were only separated in the 1880s with the growing focus on irrigation for agriculture.

Stuart Murray represents better than almost anyone the links between mining water systems and the development of public water infrastructure. Born in Dundee in 1837, he came to the Victorian goldfields in 1855 and qualified as mining surveyor, architect and civil engineer, becoming a government surveyor for mining settlements in northern Victoria. In these years he acquired valuable knowledge of Victoria's rural geography and the vital importance of water. His entry in the *Australian Dictionary of Biography* reveals that his dedication to water conservation was inspired by the cry of a settler's child for water in a dry summer. In the 1880s Murray left mining and turned his surveying skills to water, becoming a high-profile hydraulic engineer and administrator. He worked closely with

Alfred Deakin and supervised major works that formed the back-bone of the colony's irrigation infrastructure, including Goulburn Weir and the associated Waranga Storage, Laanecoorie Weir, the Upper Coliban reservoir and several works on the Murray River. In 1906 he served as inaugural chairman of the State Rivers and Water Supply Commission.[19] As this curriculum vitae indicates, irrigation and water supply had deep roots in the mining industry.

The miners who learned to read the Australian environment to successfully capture and move water had another, more surprising legacy, one that continues to touch home gardeners and farmers around the world.[20] In the 1930s a young man named Percy Yeomans was employed as an assayer and inspector of alluvial gold and tin mines in Victoria. He travelled around the bush talking to miners working small claims, men encouraged by the government's sustenance, or 'susso', scheme to work these small sites during the Depression. As Yeomans travelled he was impressed by the importance of water. He noted that successful miners had built dams to store water.[21] Building dams and moving water along the contours of hills enabled the riches of the land to be utilised. He also noted the eroded gullies of the goldfields and understood the importance of soil conservation, especially for farmers. Yeomans wondered if the water systems developed by goldminers could be applied to agriculture, bringing farmers the same benefits. He went on to apply his ideas on rural properties in New South Wales, eventually developing the 'keyline' system of water management for farming. Keyline established principles for water control by contour ploughing, channel excavation and the strategic location of farm dams. The system aims to maintain water in the landscape and promote soil fertility. Yeoman's ideas spread around the world after the Second World War, becoming one of many approaches shared by a new breed of agricultural scientists and educators working to increase global food production. Closer to home, David Holmgren and Bill Mollison incorporated keyline water management principles into the system of landscape design known as permaculture.[22] Through Yeomans and those he influenced, the water expertise of miners continues to flow through the land today.

As we've seen, the entrepreneurs who built new water networks also redefined water as a commodity that could be bought and sold. In the English common-law tradition, riparian rights entitled adjacent landholders to the reasonable use of water, so long as they didn't appreciably diminish its flow; and importantly, they didn't own the water itself.[23] Victoria's water bosses changed that. They leased the right to extract water, usually at minimal cost, and then disposed of that water for a profit. Men like John Reilly made fortunes selling water. The misguided investors in the Lal Lal Waterworks System lost thousands of pounds because they couldn't distribute and sell enough water to recoup their costs. The commodification of water required regulation to establish a stable framework for the market to operate. Investors needed the security offered by regulation – confidence that the capital spent on infrastructure had a legal basis.

The new system that emerged to regulate access to water in Victoria established several principles that continue to shape Australia's relationship with water today. One was the ongoing public ownership of waterways. The government was prepared to allow water bosses to sell water they had captured and stored, but they were not going to surrender the land through which it flowed. The principle of maintaining public easements along navigable waterways was established in the 1850s and extended under the Land Acts of the 1860s.[24] When laws about water rights were finally clarified in the *Mines Amending Act* of 1862 and the 1865 *Mining Statute,* provisions included the survey of a strip of land up to thirty-three feet (ten metres) wide along the artificial channels built to carry the water. Later, when the surrounding land was sold, these water races remained as Crown reserves. As Crown land was gradually surveyed and sold from the 1860s, this principle of reserving the banks was extended to natural creeks and rivers. Portions up to thirty metres wide along waterways were held back from sale and retained in 'the public interest'.[25] Access to streams and lakes in Britain had long been restricted by private ownership, but settlers who came to Victoria had a deep emotional desire to keep water frontages in public hands.[26] This process was initially an administrative procedure in the Lands Department but it was formalised by Order in Council in 1881. At the stroke of a pen, all unalienated land within twenty to thirty metres of the colony's rivers, creeks, aqueducts, lakes, reservoirs, swamps and inlets was reserved.[27] These strips were often incorporated into private land holdings, but nevertheless it meant that, legally at least, streams flowed between narrow threads of land separated by their own boundaries from the properties that surround them.

In effect, these land reforms created a situation in which the public continues to own the narrow strip of land alongside most waterways in Victoria.[28] Crown water frontages are only about 0.4 per cent of the total public land in Victoria but they are spread across the state, protecting an invaluable public resource. Waterways within these reserves include not only natural streams and rivers but also the artificial channels built by goldminers. John Pund's race is one of them. Although the channel itself is now fragmented and incomplete, it still sits within its envelope of land marked on modern cadastral plans and carries water from the springs at Stanley. Municipal council staff continue to wrestle with managing a network of Crown reserves or easements that criss-cross land that is mostly now privately owned. This provides Victorians with a unique opportunity for the holistic and integrated management of waterways all along the state's catchments. In some parts of the state, Crown water frontages are incorporated into broader reserves such as state forests, but elsewhere they may be the only land that remains in public hands.

Frontages make a significant contribution to the wellbeing of the landscape. They retain vital patches of remnant pre-European vegetation and are a haven for endangered wildlife. Since the 1990s, they have been managed by a network of catchment management authorities, which monitor the health of the land, water, plants and animals in their catchments, working with landcare groups and primary producers to promote biodiversity, reduce soil erosion and carry out revegetation. Frontages are not always as publicly accessible as they could be, particularly where they're enclosed by privately held land, and like other resources held in common they can fall victim to neglect.[29] Imperfect as these frontages may be, the retention of this land as a public asset provides vital opportunities not only for ecological renewal but also for cultural renewal and reconciliation. Under the *Traditional Owner Settlement Act* 2010, frontages, like other parcels of Crown land, are subject to settlement agreements with traditional owners, and catchment management authorities are increasingly working with Aboriginal people in the joint management of waterways.

Water supply for mining provided a foundation for the system of water rights that continues to shape Australian life today. Access to water via allocations and entitlements is the basis for primary production throughout the Murray–Darling Basin, a vast region drained by the Murray and Darling rivers and their tributaries. Debates about access to water in the Basin occur in the marketplace, where allocations and entitlements can now be traded, transferred, leased, amalgamated, subdivided, bequeathed or mortgaged.[30] How water is allocated between users is still highly contested, but Australia's integrated national framework is globally distinctive and underpinned by the assumption that water is a shared public good.

The Murray–Darling Basin encompasses most of New South Wales, all of the Australian Capital Territory, half of Victoria, a good chunk of Queensland and a significant portion of South Australia. It extends from the mountains near Canberra into the cattle country of central Queensland, and its western fringes stretch to Wilcannia in western New South Wales and the dry shores of Lake Mungo before it reaches the sea south of Adelaide. Altogether it covers one-seventh of continental Australia or about one million square kilometres. The Basin produces up to seventy-five per cent of the nation's food, along with cotton, rice, grain, dairy and meat for export.[31] It also hosts Australia's most ambitious water scheme, built between 1949 and 1974 – the Snowy Mountain Hydro system, which diverts huge volumes of water down the Murray and Murrumbidgee rivers. Water access across the Basin is managed by state legislation, with reference to the Murray–Darling Basin Plan, introduced in 2012. The Plan is an agreement between state and federal governments designed to allocate shares of water according to estimates of how much is in the system and expected rainfall. Alongside farmers, municipalities and the mining industry, First Nations groups are among the stakeholders, with water entitlements ('cultural flows') intended to improve the social, spiritual, environmental and economic conditions of Indigenous peoples.[32]

The Australian water market is one of the most well-developed water markets in the world, worth billions of dollars annually. It is a critical part of the agricultural industry and has spawned a whole sub-industry of economists, traders and policy advisors.[33] Private water traders hold enormous volumes of water within the system. In 2016–2017, over 7000 gigalitres of water were traded, most of that in the Southern Basin, which covers Victoria and south-eastern New South Wales. For water to be traded in such a competitive market, states legislation has operated to separate water access rights from ownership or occupation of land. This began in 2004 with the Council of Australian Governments' National Water Initiative, which has reoriented the legal principles of water ownership so that properties adjoining rivers and streams no longer have automatic rights of access to water if those rights were sold by a previous owner. It's a fundamental shift from the riparian principle of English customary law and it has been a lengthy process to get to this point. Legal scholars increasingly recognise a broad overlap between the public and private spheres of water control in Australia, where governance and regulation meet the privatisation, commercialisation and commodification of water.[34]

Today's Murray Darling Basin Authority is the institutional descendent of the original River Murray Commission established after Federation.[35] The commission brought together representatives from Victoria, New South Wales, South Australia and the new federal government to develop, from 1915, the first coordinated approach to managing the river on a national scale. Victoria played a leading role in the River Murray Commission due to its experience with irrigation agriculture. By the twentieth century, Victorian irrigators were diverting water from the Goulburn and Loddon rivers, both tributaries of the Murray, via weirs and channels built at public expense. Victoria was a leader in the development of water resources, the first to see the possibilities for irrigation and the first to legislate control over waters.[36] Early Victorian schemes provided models for the much more extensive irrigation networks introduced under the Commission, and an important feature of Victorian experience was public ownership of water.

As the second prime minister of Australia, Alfred Deakin was influential in shaping water policy at a federal level. Deakin's father had started out in Melbourne in the 1850s as a water carter, so his son understood how valuable it was.[37] As minister for water supply in the 1880s Deakin was responsible for formally nationalising water and bringing it under the control of the state. Clause 4 of Deakin's *Irrigation Act* 1886 specified that 'The right to the use of all water at any time ... [shall] ... be deemed to be vested in the Crown'. Victoria was the first Australian jurisdiction to take this momentous step.[38] Deakin's irrigation schemes usually dominate stories about water management in Australia, which typically begin with squatters and leapfrog the intricacies of mining water systems.[39]

But Victoria's attitudes to public water began long before Deakin, in the mountains of the central goldfields. Water bosses who sold water to goldminers made a healthy profit from the commodity they controlled, but because they didn't extract water from their own land – that is, they weren't riparian owners – they had no actual rights to use water under the traditional legal systems with which they may have been familiar. While they were prepared to spend large sums to secure a supply, they also wanted clear legal protection of their investment. As district Mining Boards and legislators worked out the rules for water regulation, one of their most important steps was to assert public ownership of the water. The government could then determine how to manage access. Thus, right from the outset, Australia's first water market was based on the principle of national-ised water unattached to land ownership. Water could be channelled far away from a stream without the need to own riparian land. By the time Deakin nationalised water for irrigation twenty years later, he acknowledged he was completing a process that miners had begun years before.[40] The billions of dollars spent buying and selling water in Australia today can be traced back to the 100 guineas a week John Reilly was earning from water sales at Beechworth in 1853.

Like the system of Crown water frontages that keep rivers in public hands, the Australian water market and the Murray–Darling Basin Plan are far from perfect. The system is not always properly enforced and there have been notorious allegations of users taking far more than their entitlement.[41] Cotton growers upstream in northern New South Wales, graziers in the Riverina, and downstream urban dwellers in Adelaide still argue about how to define a fair share. Governments also struggle to figure out how best to make provision for environmental and cultural flows, with no agreement on where in the 'queue' they should be, and compelling arguments that Indigenous water is the starting point after which other uses should be considered.[42]

Another major problem is groundwater, which remains inadequately understood. Stanley, near Beechworth and the site of Victoria's first commercialised water use, is today the focus of a struggle to control groundwater. As we saw in Chapter 3, John Pund and his neighbours got much of the water they sold from tunnels that tapped into underground springs. The flows were abundant and reliable for much of the year. Now those same springs are being targeted by a new generation of water bosses who want to extract the water, truck it away to a factory, and bottle it for sale in urban shops and cafes. The local community of orchardists and market gardeners, who rely on streams and dams for horticultural production, are deeply concerned about the effect this will have on local groundwater and surface supplies; they fear that removing groundwater will disrupt surface flows. The dispute has slogged through the legal system for years, highlighting gaps in local planning schemes, a general lack of scientific understanding about groundwater reserves and the under-regulation of groundwater licences. In December 2016, the Victorian Supreme Court upheld the right of the water company for bulk extraction under its licence, but the community has vowed to fight on.[43] As the use of groundwater increases around the country it appears that Stanley, a place once at the centre of water disputes, will again be at the vanguard of changing attitudes to water regulation.

Notwithstanding these ongoing problems, Australia is unusual in the world for having any kind of national system for allocating water and managing shared waterways. Surface and groundwater flowing across state boundaries demands cooperation to manage sources sustainably. The system provides a model for other countries. One of the biggest barriers faced in many places is private riparian rights of access.[44] In Australia we argue about who gets what, but there is nevertheless a common understanding about the rules and regulations and how the process will be negotiated. There is also a common understanding that water itself is a public good, not simply a private right, and that the states play a role in ensuring that water benefits more than just those who are its immediate users. In short, we have the tools available to manage our water along with the communities and environments it sustains in the best way possible, and it is up to us to make use of them.

Another legacy of the sludge epoch is the body of regulation governing the environmental responsibilities of mining companies around Australia, many of which have their origins in those initiated by the Sludge Abatement Board. As we noted above, for approval to operate a mine in Victoria today, companies must undergo an environmental assessment process and prepare a works plan that covers the life of the mine from exploration to decommissioning.[45] Plans must include provisions for dealing with a range of operational impacts, including vegetation to screen the visual effect of workings, methods to control dust, runoff and erosion, traffic management, and plans for the stockpiling of soil. Miners require a licence from the local water authority for any extra water they might require, and they must ensure that the water they use will not have a detrimental effect on surface and groundwater supplies for other users. Before work even starts, operators must provide for the eventual remediation of the site, with the goal of returning it as nearly as possible to its previous use and condition.[46] They need to strip and store topsoil so its precious cargo of local seeds and organic material can be spread out to revegetate the surface when the mine is finished. They need plans to remove buildings and equipment and to carry out ground-works for restoring the contours of the land, as well as contingency plans to control acid mine drainage. Tailings are now taken so seri-ously that Victoria has detailed guidelines for how mining waste will be managed.[47] Companies must build tailings dams to store process-ing wastes on site and construct drainage networks to isolate tailings from floodwaters, and they need to explain how they will leave those dams dry, stable and revegetated for the long term with any harmful chemicals sealed from contact with air and water.

Environmental regulation isn't confined to Victoria. It can be found in all states and territories. Laws in New South Wales and Queensland are even more stringent, requiring environmental approval before a mining lease is granted.[48] Guidelines for best practice are promoted by peak industry groups such as the Minerals Council of Australia and the Australian Institute for Mining and Metallurgy. Environmentally sustainable mining is now a thriv-ing area of research and debate, with roundtable groups, working parties, university centres, Life of Mine conferences and industry journals.[49] The series of best-practice handbooks launched in 2006 by the federal government was a joint initiative of the Department of Industry, Innovation and Science and the Department of Foreign Affairs and Trade.

Around the world, tailings dams are now the standard approach for dealing with liquid mining waste. Australian best-practice handbooks have been translated into eight languages, including Mandarin, Spanish, Indonesian and Mongolian.[50] The country's leadership in this area is due in no small part to Victorian min-ing. The system of warden's courts that developed in Victoria from

1858, a tribunal positioned beneath the Court of Mines in each district that heard boundary disputes and infringement cases, spread throughout Australia and to New Zealand, Papua and New Guinea, Malaya, and to New Brunswick and Nova Scotia in Canada.[51] The 1865 *Mining Statute* provided a template for other colonial governments to develop mining legislation based on the Victorian experience, adapted to suit local conditions. In 1907, the NSW Sludge Abatement Board was set up with almost identical responsibilities to its Victorian counterpart, and the first environmental provision in Queensland's mining law related to sludge and was modelled directly on Victoria's 1904 *Mines Act*. This 1989 legislation was the first in Queensland to protect water quality. The Sludge Abatement Board was the first organisation in Victoria charged with controlling soil loss and became the precursor to the Soil Conservation Board (later Authority), formed in 1940, itself an influential conservation agency.[52]

Victoria's expertise in managing sludge continued to play an important role in the 1970s in the establishment of the Environment Protection Authority. The first agency of its kind in the world, the Authority aims to maintain a healthy environment for the whole community. Rather than set arbitrary pollution controls on industry, the Authority was founded on the principle that the environment is public property – discharging waste into water, air and soil is not a right but a privilege that the polluter has to pay for. *Preventing* (rather than 'abatement') hereby become the new goal.[53] Inaugural members of the Environment Protection Authority Council included representatives from the State Rivers and Water Supply Commission, the Soil Conservation Authority and the Mines Department.[54] When Western Australia passed legislation to set up its environmental management system in 1971 it used Victoria's Environmental Protection Authority as a model.[55]

Many of these environmental controls have their origins in the work of the Sludge Abatement Board. Victoria's current policy document on the construction of tailings dams can be traced back to information provided to miners in the Board's annual report of 1906, which included detailed instructions and sketches for sludge dams. The information was made available to recalcitrant hill sluicers like John Pund and his son Percy when the Board began compelling them to contain their waste in settling ponds in 1909. Percy Pund's sludge dams are still part of the Three Mile Creek valley today.

The Sludge Abatement Board's work on bucket dredges pioneered techniques to restore devastated land to its former condition. This was a turning point in mine management that marked the beginning of today's 'life of mine' approach. The Board advocated re-soiling to promote plant growth and published 'before' and 'after' photos of dredged land: 'before' photos emphasising the pock-marked wastelands of old diggings that the dredges sometimes reworked; 'after' photos featuring proud farmers standing in crops of lucerne and new orchards planted on dredged ground. The Board reported extensively on operations like the Briseis Company, which began trialling new methods in 1909 for stripping and stock-piling topsoil ahead of dredging. Like tailings dams, these methods are now routinely expected of new licence holders.

Environmental provisions – both in mining law and in practice – do have many flaws. Rehabilitation requirements are not consistent between states – or even within states – and are not stringently followed.[56] The Environmental Defenders Office, a community legal service, points to a number of failings in the way that Victoria deals with the regulation of mining today. The Office has pointed out a lack of transparency and rigour in the preparation of Environmental Effects Statements, too high a level of government and ministerial discretion, the ability of mines to operate without planning permits, and the lax and ineffective enforcement of breaches when they occur. In 2012, the Office could still make the claim that the mining industry 'holds a privileged position in Victorian law ... [It] is exempt from, or receives privileged treatment under, a wide range of regulatory requirements in Victoria. Mining is treated more favourably than other activities.'[57]

Tailings dams can fail spectacularly when they overflow or collapse. A storm in 2013 caused flooding and overflow of heavily polluted waste water at the abandoned Mount Morgan mine in Queensland.[58] The disasters in Brazil in 2015 and 2019 occurred because tailings dams at two iron ore mines failed catastrophically. Numerous such dams have failed worldwide in recent decades, releasing massive cocktails of water, mud and toxic materials into the environment, and dams that are insufficiently lined or maintained can leach contaminants into nearby waterways.

When tireless farmers like Hopton Nolan at Tarrawingee and Alexander Clarke at Newstead made efforts to keep the sludge problem at the centre of public debate, they were part of the tradition maintained today by groups such as the Mineral Policy Institute, the Environmental Defenders Office and 'Lock the Gate', a grassroots movement that opposes fracking and mining on farmland.[59] In general, there is now a broad consensus among regulators and the industry that the environment must be taken into consideration. Community attitudes have shifted and miners now need a social licence to operate. Industry awareness is evident on websites where images of plants, native animals and scenic landscapes appear alongside photos of dusty miners standing proudly next to enormous machines. It is also evident in the press releases quickly distributed after incidents, such as the discharge of tailings from the Newcrest mine near Orange in February 2018 or from the Fosterville

Gold Mine near Bendigo a few months later.[60] Community scrutiny continues to play a vital role in keeping the industry honest. Healthy scepticism and critical vigilance remain necessary, but the media releases of today's miners are a long way from the dismissive response to Jacques Bladier's complaints when his vineyards were destroyed by the Bendigo puddlers in 1859.

Part of remembering the origins of modern environmental regulations is to more fully appreciate the unsung heroes of the mining era, those whose persistent agitation eventually resulted in sludge control. Community groups that fought the sludge, like the Kiewa Anti-Dredging League, the Loddon Anti-Sludge Association and the Ovens River Frontage Anti-Sludge Pollution Association, are overlooked in the history of the environmental movement, but deserve to be remembered. Farmers like Hopton Nolan and Alexander Clarke and politicians like John Bowser and Henry Parfitt are some of our first environmental warriors, even if they wouldn't have recognised that label themselves.

Nolan and Clarke, for example, were public-spirited farmers who used their leadership skills to mobilise their neighbours and established local action groups to give weight to their cause. When it came to protesting the destruction of farmland they knew the power of political lobbying and the value of collective action. Nolan in particular played a prominent role in the anti-sludge campaign through his defence of the Ovens River over several decades. He was an active member of the local committee agitating for a sludge channel in the 1870s and in 1886 took the visiting members of the Sludge Inquiry on a tour of the district to see the damage. At the age of seventy-nine he was again giving evidence when the Sludge Abatement Board visited the district.[61]

Hoppie's neighbour Henry Parfitt was the lone voice in parliament calling for the tabling of the recommendations made by the

Board Appointed to Inquire into the Sludge Question in 1886. He was unable to solve the problems, but one of his successors as local representative was John Bowser, who, as we know, did more than anyone to protect Victoria's rivers. Bowser's entry in the *Australian Dictionary of Biography* doesn't mention sludge or water quality but these were causes he championed passionately.[62]

Various other politicians played significant roles. Both Ewen Cameron and Donald McLeod had reasons to be affiliated with the mining cause and yet they repeatedly advocated for clean water. As ministers – for mines and water supply and mines and forests respectively – they sought to include sludge control provisions in the Mines Acts and then steered them successfully through parliament. Though Cameron represented Warrandyte miners for years on the Castlemaine Mining Board, it was the sludge oozing into his beloved Yarra River that inspired him to act. He told parliament of the thick sludge in the river, and of his visits to 'try to persuade the miners to prevent so much sludge running into the river'. 'I urged them strongly', he lamented; but Cameron ultimately saw that 'persuasion' and 'urging' weren't enough.[63] When he introduced the sludge provisions in the failed mining bill of 1903 he was consciously moving against his fellow mining members in order to defend the rivers. McLeod, Cameron's successor, was also a member from a mining area, but he also valued water, and he worked with John Bowser to get the sludge provisions passed in 1904.

Modern environmental sensibilities were in their infancy at the start of the twentieth century when McLeod and Bowser created the Sludge Abatement Board, but there are hints of increasing awareness in the archive. Elwood Mead, a witness quoted in the 1913–1914 Board of Inquiry into Complaints of Dredging and Sluicing, said of the Ovens Valley that:

those mountains form one of the great health and pleasure
resorts of this State ... the number of people who go there for
the summer will very greatly increase and it is becoming a matter of both aesthetic and monetary value, and it would be better
to have the valley looking as good as nature made it in the first
place ... it would be a misfortune – in fact, a calamity – to have
the agricultural land of that valley destroyed.[64]

Mead was a prominent Californian brought to Victoria in 1907 to
succeed Stuart Murray as head of the State Rivers and Water Supply
Commission. He had worked as an irrigation expert in Colorado and
Wyoming before his Australian appointment, and when he eventually returned to the United States he became federal commissioner
for reclamation. He was a charismatic figure who did much to shape
Australia's irrigation agriculture industry.[65] For the most part, Mead
was concerned with the impact of sludge and dredging on agriculture, but his greatest eloquence was reserved for the natural beauty
of the Ovens Valley and the value of the environment. Although the
1913–1914 Board of Inquiry collected testimony from witnesses all
over the state, that testimony didn't make the final report, but Mead
was considered worth quoting at length.

Mead extolled the natural beauty of the Ovens because of its
potential as the basis of an emerging tourism industry. At that time,
tourism was in its infancy, but it was becoming increasingly important, particularly in the mountainous north-east of Victoria. The
Mount Buffalo Chalet was established in 1910, and from the early
years of the twentieth century there was a concerted push to bring
tourists to Beechworth.[66] Even Alfred Billson, staunch defender of
sluicing, dredging and mining generally, proclaimed the virtues of
the region as a place of beauty and a resort for health.[67] In fact, Billson
became one of the great local tourism promoters, the irony of which

was pointed out in parliament by the indefatigable John Bowser.

Most politicians adopted an instrumentalist perspective on the environmental issue: it was economics that mattered. Control of sludge was tacit acknowledgement of the increasing value of farming, especially irrigation agriculture, and the waning power of mining. Legislators responded with enlightened self-interest, and parliamentary debates reflected their view of nature. Arguments were framed in terms of the relative economic benefits of agricultural production and goldmines, the numbers employed in each, potential future incomes, and the cost of infrastructure already built. Neither politicians nor community groups specifically invoked what we would consider today the values of 'conservation'. There was, for example, no discussion of biodiversity or species loss, and insofar as the ecosystems of the rivers were considered it was solely in terms of recreational fishing.

Fish may have gotten short shrift, but several figures were aware that more was at stake than economic prospects. Bowser was conscious of future generations:

> What view will the young people of the future make of us when they look back? What will they say of us if we have allowed the life-giving Ovens and other streams to be destroyed when they should have been preserved for posterity? Surely that is a grave trust reposed in honorable members and in the Government of this State.[68]

Members of the 1913–1914 Dredging and Sluicing Inquiry Board also took the long view. More dredging, they reported, would result 'in the widespread destruction of large tracts of magnificent country … which should be the heritage of the future'.[69]

These considerations of the broader dimensions of industrial processes reflect not only the changing economics of mining but also shifting social values and the emergence of an identifiable conservation movement. The years between the Sludge Inquiry in 1886 and the passage of the *Mines Act* 1904 saw a number of significant steps towards environmental protection.[70] The first national parks were established in New South Wales and Victoria in the late nineteenth century, including Mount Buffalo National Park near Beechworth in 1898, and enthusiasts joined natural history and bushwalking clubs to share their enjoyment of the natural world. The Australian Natives' Association (an organisation for Australian-born whites) championed the establishment of national parks and lobbied on behalf of the anti-sludge movement. Melbourne newspapers ran a strong campaign against dredging. When Bowser, Cameron and McLeod spoke on behalf of rivers they were reaching a public increasingly exposed and receptive to such arguments.

Groups like the anti-sludge leagues in the Kiewa, Loddon and Ovens valleys deserve greater recognition for their contribution to the environmental movement. They were among the first grassroots, community-led organisations to call for action to protect the environment, and they had statewide influence. And these were only the most visible and best organised among many community groups and concerned citizens around Victoria. Together these people represented a groundswell of public opinion. The health of the rivers may have been a by-product of their concerns, but it was nonetheless a real consequence and beneficiary of their work. The value of their efforts was acknowledged in the 1930s by a new generation of landholders fighting a revival in the dredging industry who hailed their 'valuable work' in stopping the damage.[71]

The anti-sludge campaign sits awkwardly in modern histories of environmental awareness. It would be many years before another environmental issue would generate such widespread interest and mobilise such strength of public feeling. In 1969 concerned citizens around Victoria rallied to protect the Little Desert in north-western Victoria from clearance for agriculture and subdivision.[72] This new generation of urban-based activists had forgotten the earlier fight for the rivers; that fight's rural rather than metropolitan networks and its language grounded in economics and industry conceal its significance. Ironically, the success of the anti-sludge campaign has also contributed to its erasure. Today we take for granted the unpolluted water that flows through Victorian rivers and rarely think to question whether it was always thus. It is time we remembered the work of those who ensured that legacy.

A final dimension of our collective forgetfulness about sludge is the rivers themselves. As inhabitants of this continent, we have largely forgotten to include mining in our historical understanding of how rivers have changed since Europeans arrived. Our laws for sharing water and managing mining leases are the practical legacies of the sludge problem. Forgetting the mining origins of these legacies is unfortunate but it's also relatively benign. It is a much more serious thing to forget what was done to the rivers themselves.

The history of our damage to them continues to affect waterways and floodplains today and will continue to do so long into the future. Rivers are front and centre of many conservation efforts but when we forget their histories we are less able to take the most appropriate actions to manage them. Because sludge has been forgotten there is little robust evidence of the impact it has had. It simply has not been considered in the body of scientific literature on Victorian rivers. Consequently, we know virtually nothing about sludge in the physical environment today. We don't know where it is, or the nature of its physical and chemical properties; we don't know what it means for farming or catchment management or revegetation or river health or erosion control. It stands to reason that all the mud that filled waterholes, buried fences and covered crops 100 to 150 years ago is still out there somewhere, and, whether we realise it or not, we still have a sludge problem.

In order to learn more about sludge today we have begun working with a group of river and soil scientists, although this work is still in its early stages and there is much more to be done.[73] One thing we have learned about already is the sheer volume of sludge that was dumped into rivers.[74] Our team in the Rivers of Gold project combined detailed analysis of data from old Mines Department reports with geospatial mapping to calculate historical sludge volumes. Based on this work we estimate that between 1851 and 1900 over 650 million cubic metres of sludge was generated by gold mining in Victoria, most of which went into creeks and rivers and out onto floodplains. Unsurprisingly, the region with the greatest volume of sludge was north-east Victoria, where ground and hydraulic sluicing were most common. Ballarat and Bendigo also produced considerable amounts from quartz crushing and puddling. We are still working to understand the effect of all this sludge on rivers. What we do know is that Victorian rivers today are nothing like those that the First Nations knew 200 years ago, and mining played a major part in that change.

Rivers are dynamic. Water renews itself every moment as it flows from source to mouth. The banks and the beds of the river change too in an intricate and ongoing dance between the moving power of water and the inertia of earth and rocks that form the channel and floodplain, with the added complexity of changing rainfall and vegetation. The interplay of water and earth creates a fluctuating cargo of sediment that is picked up, carried and redeposited as water moves downstream. The sediment load joins this choreography, helping to reshape the river as the particles move between bank, bed and floodplain. Rivers constantly seek equilibrium between the gradient of the channel and the rate of water flow within it. On steep slopes the water flows quickly. It has the energy to erode sediment from the channel, eventually reducing the gradient and slowing down the flow. On flatter ground the water slows down and loses energy. It can no longer carry as much sediment, so it drops its cargo and the riverbed gradually rises. On level valley bottoms the river repeatedly meanders back and forth across its floodplain, carving out sediment from the outside margin of bends and depositing it again on the inside of the curve.

Rivers use sediment as a tool to adjust their paths. Rapid flows carry more sediment and larger particles such as sand and gravel. Powerful floods can even move boulders. Slow, languid currents carry only tiny particles of clay and silt. Narrow channels concentrate the power of the current, increase the speed of flow and carry more sediment. If there is not enough sediment already in the system, the river seeks more by eroding the bed and banks. It gradually straightens out curves in the channel and incises the bed deeper and wider. Where the channel spreads out, the force of the current is dissipated, and the sediment is dropped. If there is too much sediment for the current to carry, then the river drops its cargo quickly, filling the bed, spilling over the floodplain, and creating a wide, braided system with multiple channels.

Sludge interrupted the normal functioning of rivers by radically increasing the cargo of sediment carried. This had both immediate and longer-term effects. Initially the rivers struggled to cope with all the extra sediment. Most of the time there was not enough water flowing to carry the load, so the sediment dropped into the riverbeds, filled up the original waterholes and eventually clogged the whole channel. John Beveridge was a grazier near Clunes who recalled waterholes on Tullaroop Creek large enough to float a ship; years later the creek was simply 'a stream of running sludge'.[75] James C. House was shire engineer at Newstead in the 1880s and remembered a large waterhole on the Loddon at the start of the gold rush that was a quarter of a mile long and up to thirty feet deep. By 1886 it was 'entirely filled with sludge', just like all the other waterholes along the river.[76] Sludge-choked rivers spread out over the floodplain, with multiple small streams depositing mining sediment across a wide area. Sometimes, storms and floods delivered enough extra water to pick up the sediment and carry it further downstream, spreading it into the valleys. Albert Bucknall observed that quartz tailings and alluvial sludge flowed down Tullaroop Creek as a matter of course, but floods spread them out all over the flats.[77] All these processes can be traced in the historical record, in the testimonies of farmers watching the rivers change.

Through the Rivers of Gold project we are trying to find out what happened next. There are clues in the knowledge gathered by river scientists and in comparative studies of river behaviour overseas.[78] In Victoria, the rivers gradually adjusted to the new sediment load and changed from their original chains of ponds and deep waterholes to continuous shallow streams that flooded easily. But then mining ended and the extra sediment load ceased. Without the sludge the water moved more quickly again and did not spread out and rid itself of cargo. Instead, it returned to a single channel and started to gather sediment from the bed and banks. In the 100 years since mining ceased, rivers have incised deep channels into the floodplains, like the one at Tarrawingee where we first saw sludge in the eroded river bank.

Changes to the Ovens River have been studied by two geography students at the University of Melbourne. Both used the valley as a case study to analyse long-term river recovery after disturbance. Written several decades apart, both studies reveal the river struggling to re-establish its former contours. The first thesis was written in 1979, only twenty-five years after the last dredge stopped work at Harrietville. Here, David Beard observed that the Ovens River downstream from Myrtleford 'has changed from a tight meandering to a straight braided stream as a result of the mining debris deposited in local watercourses upstream. The surplus energy created by this change in stream characteristics has led to increased erosion of the downstream section while further upstream the river remains stable.'[79]

When Deborah Cargill returned to the valley in 2005 she found that little had changed.[80] The river below Myrtleford was still straighter, less sinuous and faster-moving than it had been before dredging. Upstream in the dredged areas the river had become locked in its channel by the spoil heaps of heavy gravels that form the river banks. It could no longer move naturally across its floodplain to create new meanders.

Some rivers are now moving faster and eroding, but in other places they are still constrained by sludge. The fine silts and clays moved downstream long ago, part of the burden of sediment accumulating in irrigation dams and weirs on the Murray.[81] The heavier sands and gravels, however, are still in the rivers. Every time a major storm brings a pulse of floodwater, these 'sand slugs' move a little further downstream. Sand slugs choke the river, creating a smoother, shallower stream bed and destroying habitat by engulfing vegetation in the stream and on the adjacent banks.[82] We have seen these on the Loddon at Newstead, and there is another on Reedy Creek below Eldorado. Sand slugs will continue to move down the rivers for generations until they reach the Murray, and, some day, the sea.

People may have forgotten about mining sludge, but the rivers remember. American geographer Ellen Wohl writes that 'Rivers reflect a continent's history ... They also reflect a people's history ... River valley sediments record all the changes in a river's drainage basin over thousands of years'.[83] Victorian rivers are an enduring archive of nineteenth-century gold mining.

*

There is much to celebrate about the gold rush and its transformation of Victoria. Its rich historical legacy includes universal male suffrage, the secret ballot, high levels of home ownership, substantial public infrastructure, cultural institutions including libraries, museums and art galleries, and an impressive array of private and public buildings. While we remember these aspects of gold we should also observe the legacy of more sustainable mining practices and the shared use of water. And we should also acknowledge the enduring cost of gold and the damage done in finding it.

We should also admit that which hasn't been remembered. It has become clear to us that much has been forgotten about Victoria's mining past and its legal, social and environmental legacies. To return to our first question: *why* have we forgotten so much? The answer, in large part, is the louder stories that continue to be told about the gold rush.

The generation who participated in the rush of the 1850s were fully aware they were making history and considered their deeds worth recording. They chronicled their travels and activities, publishing them as letters, diaries, reports and images.[84] There were guides for emigrants to life on the goldfields. Some writers, like English journalist William Howitt, were effectively on assignment, providing regular updates from the front to be published in periodicals for a wide readership back 'home'. Others, like James Robertson, building water races in the Creswick forest, wrote private letters to parents knowing they would be shared between family members and neighbours and possibly even published in the local newspaper. Still more, like Edward Snell and Seweryn Korzelinski, wrote diaries for posterity – for descendants they hoped would be interested in their ancestors' adventures. As this generation aged they crafted memoirs that captured the voices of pioneers before they passed.[85] Others devised broader narratives, casting the events of their lives within a larger social context. In 1870, William Withers looked back on nearly twenty years at Ballarat. He'd seen many changes and could see more coming, so he wrote the first history of the town, 'to gather some of the honey of fact from fugitive opportunity, that it might be garnered for the historian of the future'.[86]

From the outset of the gold–rush era, Victorians very self-consciously historicised those early years of excitement and discovery. These firsthand accounts share an optimistic, celebratory tone that was then echoed by scholars in the twentieth century. Hardship and tragedy are presented as obstacles on the path to personal and collective triumph; forbidding wilderness gave way to a settled, prosperous, civilised society. Mining provided the wealth that facilitated achievement, even if it was only a modest stake in a business or a home secured with a miner's right. Mining also provided the wealth that underpinned civic pride in goldfields towns and in Melbourne itself. It built imposing town halls, court houses and post offices, paid for clean water supplies and railway networks. Mining wealth built mansions and endowed churches, mechanics institutes, schools of mines, art galleries, botanic gardens and public art. William Withers at Ballarat, looking back on a settlement that two decades previously was only 'wild bush' and now had a population larger than the English towns of Winchester, Salisbury, Canterbury and Lichfield combined, wrote that 'For the good done, and for the doers of the good, we may all be thankful, if not proud . . .'[87]

Later scholarship followed a similarly romantic vein. The excitement of the early rushes suffuses works from Geoffrey Blainey's landmark 1963 history of Australian mining, *The Rush that Never Ended*, to more recent treatments.[88] The late Weston Bate, a respected historian of Victorian gold, described as 'legendary' its transformative effect on the colony's population, economy, institutions and culture, and so it was, but in broader histories of Victoria, the rushes and disruptions of the 1850s serve as an interlude before the focus moves on to other issues such as closer settlement, the growth of Melbourne, and the development of irrigation and manufacturing.[89]

We have not told or been told stories about gold that present it as an industry with multiple dimensions, including technological change, regulation, commercial structures and working conditions. Remarkably, there are numerous social histories of mining in Victoria, but there is no general history of the industry comparable to those produced about other states.[90] General histories of Australian mining since Blainey's have taken a similar approach, consigning Victoria to a footnote about the 1850s before describing the industry elsewhere in the country, with only fleeting mention of the industry's longer history.[91] There have been accounts of individual companies and specific issues such as mining technology, regulatory arrangements and economic structures, and every goldfields town has its local history, but there has been no synthesis of these individual stories.[92] Urban histories of Ballarat and Bendigo present the most coherent accounts of mining's emergence as a large-scale industry.[93] Its maturation was central to the permanence and prosperity of these major regional centres, and its decline likewise shaped their communities. These histories all contain substantial hints about mining's trajectory in Victoria, but the story remains fragmentary and piecemeal.

Gold history is embodied in museum displays such as Sovereign Hill and the Gold Museum, Ballarat's premier tourist destination, but the living history park confines its portrayal of mining to the rush period from 1851 to 1861. It struggles to present more complex stories of the development of mining as an industry. While the thousands of tourists who visit each year can witness an industrial-scale array of working stamp batteries, with their attendant noise, no rock is crushed, so both the dust and tailings are harder to imagine.

Affection for the architectural heritage of Marvellous Melbourne is a less conscious but nonetheless very real celebration of the gold rush. The Royal Exhibition Building in Carlton, the first place in Victoria to be included on UNESCO's World Heritage List, was built with public money at a time when mining wealth was flooding into public coffers and the private pockets of metropolitan investors. It housed the Melbourne International Exhibition in 1881, which showcased Victorian wealth and progress to an international audience.[94] Many of Melbourne's treasured stately homes are likewise the product of gold-rush wealth. Rippon Lea Estate, for example, was built by goldfields merchant Frederick Sargood, while Quat Quatta was home to mining magnate John Alston Wallace. Fortuna Villa at Bendigo was the home of George Lansell, 'the Quartz King'. The public face of gold's prosperity is renowned internationally, but the diggings that produced that wealth haven't achieved equal status. The Castlemaine Diggings National Heritage Park is on the federal government's national heritage list, but proposals to nominate the goldfields for world heritage status remain desultory after more than twenty years of discussion.[95] In any event, neither the official citation on the national heritage list nor the most recent proposal for world heritage nomination include any reference to the environmental effects of historical gold mining.

The popular history of gold in Victoria remains associated with a triumphalist narrative of nineteenth-century progress. It has been regarded almost universally – and appropriately – as the driver of regional and colonial growth. This is predominantly a story of colour and romance rather than the harsher realities of industrialised mineral extraction, racism, brutal working conditions and fatalities. Cracks are starting to appear in this narrative, however, and other stories of gold are increasingly being told. One of the first critical voices was historian David Goodman, who in 1994 pointed to the ambivalence and disquiet felt by many Victorians during the gold-rush years. Since then, other collections have focused on Aboriginal and Chinese perspectives, and the experiences of women, and these themes have been taken up by subsequent research.[96] Clare Wright's award-winning book on the Eureka uprising repopulates the gold rush with the voices and experiences of women. There are now also substantial histories of Aboriginal people and mining. The Dja Dja Wurrung clans around Bendigo and Castlemaine have prepared a plan for their country that is a powerful critique of gold's environmental effects. They say that mining turned their land into 'upside-down country'. Several recent studies have canvassed the effects of mining on the natural environment, while the creative arts are providing new critical perspectives as well.[97]

Our dominant cultural narratives of the gold rush are not the only reason sludge has been forgotten – the land itself has colluded to help us forget. River valleys simply don't look like industrial landscapes, even if scars of old mines are still commonplace. For the most part they are green and vegetated and look reasonably healthy. Experts know that the vegetation is often from invasive species and the streambanks are eroding, but even ecologists don't see the sludge lying just beneath the surface. It is hard to imagine that these now pleasant places were once the scenes of desolation described again and again by nineteenth-century observers.

Victoria's rivers are not 'restored', because they are not and never will be what they were before mining, but they have recovered some of their ecological and aesthetic qualities, and that should be celebrated. This is in part due to ongoing riparian processes that gradually move sediment through the system. In the decades since Percy Pund built his settling dams, the dynamic processes common to all rivers have continued their work. Streams carved wider and deeper channels into their floodplains, and regular flooding declined. When major floods did occur they no longer carried mining waste across the plains. Without this constant disturbance the hills and riverbanks slowly revegetated. Rivers have been supported in their recovery by the actions of government agencies and local communities. The Sludge Abatement Board continued its work into the 1970s. Organisations including first the Soil Conservation Authority and then Catchment Management Authorities continue working to restore degraded mining areas and farmland and prevent soil erosion. Local groups have worked to restore creek and river valleys, planting native trees, shrubs and grasses, clearing rubbish and removing weeds. In many goldfields towns the rivers are no longer abandoned wastelands. Instead, they have become valued, open public space, with walking tracks and sporting grounds. They are places where small children can play in the clear water of the streams.

CONCLUSION

THE SLUDGE PROBLEM TAUGHT VICTORIANS A GREAT
deal. The new arrivals learned how to manage water in their
new home. They learned about rainfall and weather patterns; how streams and rivers responded to wet winters, dry summers and sudden intense storms. They learned that if they wanted water they needed to catch rain when it fell and store it for later use. Learning to store water forced them to look closely at the topography of the country. They surveyed the valleys and ridges of the high catchments and made maps of the routes that water could take. Through trial and error they worked out the best places to build dams in order to store water over the dry summers. They found springs high in the mountains that would provide water all year. Miners developed valuable skills and experience in holding water and moving it around the landscape.

Learning to control the flow of water also taught miners how to acquire legal control. It was not an easy or peaceful process, but eventually they worked out ways to share water among competing users. The society of which the miners were a part determined that water belonged to the Crown, which in this new society came to mean everyone. Water was not privately owned but it could be leased, diverted and used for fixed periods. Miners wanted to sell water, so they learned how to set a price that the market would tolerate. They created an economy in which water could be bought and sold as a commodity.

Once miners began controlling water, their society had to learn another set of skills: how to control the damage that water could do. Providing abundant water when and where it was needed created the problem of waste. Industrial pollution on this scale was unprecedented in the new colony, and the old ways of dealing with damage didn't seem to work. Engineering skills moved the waste – the sludge – away from some people's activities but caused more problems for others. The community and its leaders eventually recognised that industries – even the most valuable industries – needed to be regulated if their quest for profits was not going to harm others. Ultimately, community agitation, legal action and parliamentary debate and process combined to produce a new set of rules that made miners take responsibility for their actions, including their waste. This was not an easy or peaceful process either. It took more than fifty years of pressure, and there were many failed attempts. Powerful industries are difficult to regulate and hard to pursue through the courts when existing laws are not fit for purpose.

As Victorians learned to minimise and prevent harm to their rivers, they also began to learn how to repair the damage that they had done. They learned how to build tailings dams that would settle the sludge out of the water. Experience taught dredging companies how to strip the topsoil and return it to cover the land again, to grow crops and trees. They learned how to re-contour the floodplains and river channels so that country damaged by mining could be made pleasant and useful again, even if it could never be restored to what it was originally.

All of these lessons have been passed on to us. We still rely on knowledge gleaned by miners, and even on their infrastructure, to provide water for homes, agriculture and industry. We draw on miners' laws to inform how we share water, and how we manage the impact of mining on the environment. We still use the Sludge

Abatement Board's techniques to store tailings and remediate mines, and we continue to learn how to do it better. More than ever we rely on access to water frontages to manage and restore river catchments.

Any tradition that continues to have value and meaning will be continually reshaped to meet new conditions. The lessons we've learned from the sludge problem are no different. Water law and mining law have progressed a long way from their origins in the sludge era. Infrastructure has grown vastly more sophisticated and extensive, and our reliance on it has also burgeoned. River valleys have changed with new uses and new generations of residents and through the ongoing cycles of the non-human world: fire and flood, drought and growth and death. Because so much has changed, we've forgotten much of where things started, and that can make us complacent. We've forgotten the struggles that initiated these legacies, and it is easy to take for granted that which has been achieved. No matter how much has transformed since the sludge problem was solved, one of its most important lessons remains unchanged: industries of all kinds require regulation so that harm to others is minimised. The vigilance of the community is still the most important tool to keep industry, officials and politicians aware of their responsibilities, and respectful of the rights of others.

Metals mining has long had a special place in Victorian history and identity. It is as important today as it was 150 years ago, but the industry has moved on from Victoria's goldfields. In the nineteenth century the mining industry was so powerful that its rights took priority over all others. Today, mining still claims a moral priority, asserting its good for society and its capacity to bear progress and even civilisation.[1] The perceived goodness of mining is so entrenched that even now it can be difficult to acknowledge other points of view, or see the value in other uses of the land. Whether or not we accept the mining industry's view of itself and its role in society, the fact remains that all of us rely on the products of mining and we use them every day. How do we reconcile this dependence with the other lasting lesson of the sludge disaster – that the damage done by mining is forever? That more than a century later, and despite the best efforts to remediate, our rivers and floodplains will never be the same? This is true of Victoria and it is just as true of all the distant places where mining is carried out today. Whatever value mining has for a society, its costs are enduring. If we are to continue mining, the challenge today is to find ways to do so sustainably.

And if metals mining is to be sustainable – not just economically but environmentally and socially – then its effects need to be reduced. There are several dimensions to this. Existing and future mines must continue to be held to account for their impact on the environment. Politicians need to enact strong regulations to control dust, restrain water use, preserve water quality, manage tailings and remediate sites after mines close. Governments need to provide adequate resources to monitor compliance and ensure regulations are observed. Researchers and industry must seek greater efficiencies in recovering metals from the ores. Some researchers now point to the economic value of materials contained within tailings themselves: pyrites, cobalt, lithium and indium are just some of the resources that are highly sought for electronics and the batteries needed for sustainable energy. It's been proposed that these materials could be recovered from tailings using new metallurgical techniques, including processing based on bacteria. Another new technique is using chemicals already in the waste water to trigger a process that soaks up contaminants and leaves behind a much smaller volume of sludge.[2]

Another step is to seek alternative sources for minerals that don't involve digging more holes. This means recycling existing goods to recover the metals that have already been mined. The ABC's *War on Waste* that screened in 2018 highlighted the increasing amounts of e-waste Australians are accumulating in the form of discarded phones, computers and digital gadgets. These could all potentially become valuable sources of metals for other uses. In some cases, the quantity of metals in old phones is actually greater than the ratio of metal to a similar volume of ore; it's been estimated that a million mobile phones could contain hundreds of kilograms of copper, silver, gold and palladium.[3] This makes our e-waste far too valuable to be simply thrown away. The higher we value the environment, the more we will value our disused consumer goods. Why would we destroy more land with new mines when our own rubbish could be mined instead?

Ultimately, if we think mining should be sustainable and we value the environment, then we are the ones with the greatest responsibility. The sludge problem was solved because individuals like Hopton Nolan and Alexander Clarke organised their neighbours and galvanised the support of the press. They could see the damage, and it motivated them to act. Modern communities still have lobbying power, but motivation can be more difficult when most of us don't see or experience the damage firsthand. But there are many roles for city dwellers far away from where mining takes place: one is to support communities that are directly affected; another is using our power as consumers. New mines are needed to produce more consumer goods because we demand more and more new things. E-waste is lost because we don't recycle our old phones and laptops appropriately. We have the power to operate more thoughtfully in our consumption of new goods, and more carefully in our discarding of old goods.

Victorians may have ultimately solved the sludge problem, but it has never actually gone away. Sludge remains in our river valleys, and the hills still bear the scars of mining. New sludge is produced by mines in other parts of the world, and other hills are dug over for minerals. The world of the gold rush is a foreign, strange and exotic place to modern Australians, but the problems it faced are much the same. We can look at the choices made and the actions taken in the past to solve those old problems, and they can help us think about the choices that we make now to solve our own. In another hundred years we will be history too, and generations hence will decide if we have dealt fairly with the future.

ACKNOWLEDGEMENTS

OUR FIRST AND GREATEST DEBT IS TO JODI TURNBULL, who has worked with us as a researcher for much of this project. Jodi is a talented field archaeologist, highly skilled in the use of GIS and a whiz at unearthing archival treasures. Silja Carruthers, Jocelyn Strickland and Lil Pearce provided much-needed archival assistance at crucial times. Together with Jodi, we have been privileged for the past few years to be a part of the 'Rivers of Gold' project. Our conversations and fieldwork with Ian Rutherfurd, Ewen Silvester, James Grove, Mark Macklin, Jude Macklin, Darren Baldwin, Lynette Peterson, Francesco Columbi and Greg Hil have been intellectually stimulating and just plain fun. They have helped generate invaluable insights into the impact of sludge on rivers, and opened our eyes to the many ways in which legacy mine waste is important in modern ecosystems. Lynette's earlier work on Bendigo's sludge problems has been particularly influential. Honours students Rowan Frawley, Luke Gunton, Jess Hardy and Kerry Hammond have each helped us add to our understanding of water supply and use on the goldfields.

Our colleagues in archaeology and mining history have shared their knowledge and unpublished reports with us for many years and have frequently taken the time to show us their research sites all over eastern Australia and New Zealand. We're particularly grateful to Peter Bell, Chris Davey, Charles Fahey, Denise Gaughwin,

Gordon Grimwade, Greg Jackman, Rob Kaufman, Justin McCarthy, the late Barry McGowan, Peter Petchey, Neville Ritchie, Jeremy Smith, Iain Stuart, Ray Supple, Andrew Swift, Dian Talbot and Jan Wegner. David Bannear, who knows more about the archaeology of mining in Victoria than anyone, has always been gracious and generous with his knowledge. In the UK, Jane Powning, Ainsley Cocks, Adam Sharpe and Colin Buck showed us around the World Heritage tin and copper mining sites of Cornwall and West Devon, while Mark Macklin and Jude Macklin introduced Susan to the legacy of mining in Wales.

Expert local historians have taken us into their homes, shared their family and local histories, and shown us sludge and water races in their areas. We're grateful to David Horner and Jeannie Lister at Castlemaine and Derek Reid at Newstead. In the north-east we're grateful to Peter and Jan Nolan, Damien Hynes, Chris Dormer and her fellow Friends of the Stanley Athenaeum. Leon Bren, Kevin Tolhurst and Don Henderson at Creswick were generous with their time and professional expertise, and in Kevin's case his spatial data on water race networks.

Our colleagues in environmental history have prompted us to think about the broader relevance of sludge and mining's legacy in contemporary Australia. Katie Holmes, Ruth Morgan, Andrea Gaynor, Cameron Muir, Emily O'Gorman, Margaret Cook and Rebecca Jones have all been stimulating role models. The La Trobe Environmental History Reading Group, including David Harris, Katie Holmes, Liz Downes, Sandro Antonello, Rachel Goldlust, Kerry Nixon, Peter Minard, Lil Pearce and Karen Twigg, has provided a forum for exploring ideas about what it means to write environmental history and how to do it. Tom Griffiths, Libby Robin and Billy Griffiths all provided invaluable advice and opportunities. Ron Southern, as always, has been the source of meaningful conversation, stimulation and insight, as well as a knowledgeable local guide to many Victorian goldfields.

Lee Godden and Rebecca Nelson have improved this book by reading chapters and sharing their expertise on environmental and mining law. Conversations with Grace Karskens over many years have shaped our understanding of the relationship between archaeology and history. More recently, she and Clare Wright have been extraordinary mentors, supporters and friends who have helped us make the transition from writing academic archaeology to public history. Both have read and commented on drafts at crucial times, and this book would simply not exist without them.

The School of Humanities and Social Sciences at La Trobe University provided funding and much-needed study leave for Susan to write. The School of Geography at the University of Melbourne provided a home on two occasions for that writing, thanks to the hospitality of Associate Professor Ian Rutherfurd. At the Australian National University, Professor Tom Griffiths, director of the Centre for Environmental History, and Professor Libby Robin of the Fenner School provided space and welcomed Susan to the community of environmental historians in Canberra. A base in the UK was provided by the School of Geography and the Centre for Planetary Health at the University of Lincoln, both under the direction of Professor Mark Macklin. Colleagues in Archaeology at La Trobe, including Anita Smith, Nicola Stern, Richard Cosgrove, David Frankel, Andy Herries, Jillian Garvey, Keir Strickland, Colin Smith, Matthew Meredith-Williams, Cristina Valdiosera, Rudy Frank, Ming Wei and Paul Penzo-Kajewski have always provided generous intellectual companionship and logistical support.

We consider ourselves particularly fortunate to have secured the interest of Black Inc. and La Trobe University Press. The team there, including Chris Feik, Dion Kagan, Kate Nash, Lauren Carta and Erin Sandiford, have been what every author dreams of when it comes to getting their book made and marketed. Dion's infectious and irrepressible enthusiasm for the project has been an ongoing delight. Not to mention the book's fabulous cover design by Regine Abos and stylish typesetting by Akiko Chan.

Many friends have stepped in to help look after our boys during fieldwork and in times of crisis, and we are deeply indebted to them. Thank you, Danielle Afif and Greg Clarke, Janine McCarthy and Rod Gray, Janine Larson and John Howe, Jutta Howe and Romano Struder, Sue Blood and Ross Dawson. Members of the Thornbury Primary book group have always been a source of companionship and stimulation, and a compelling reminder of the audience every book needs. Our families are always encouraging and supportive, and we're grateful to Joan Lawrence, Noel Roberts, Michael and Mary Davies, and Bronwyn and Ross Hardy. It's a great sadness that our parents Muriel and Evan Davies and Marjorie Lawrence have not lived to see this book completed, but we present it to John Lawrence with love. Our children, William and James, have spent their lives being dragged around goldfields towns and searching the bush for elusive water races. We hope they've enjoyed at least some of it – if nothing else, it will give them stories with which to bore their own children.

Much of the research for this project was funded by grants from the Australian Research Council. We are endlessly grateful for its support and the support of the Australian taxpayers that fund it in turn. We deeply hope the results will repay your trust and sufficiently demonstrate the value of humanities research.

PICTURE CREDITS

Peter Davies: Map of Victoria's goldfields © 2019 (p. 16); Ian Rutherford and sludge in Hodgson Creek, Tarrawingee © 2016 (picture section); Sluicing devastation at Pink Cliffs Reserve, Heathcote © 2015 (picture section); Garfield waterwheel stone foundations © 2013 (picture section); Eatons Dam wall © 2011 (p. 75); Plan of John Pund's water races © 2019 (p. 84); Creswick water race (picture section); Siphon pipe near Fryerstown © 2013 (picture section); Pound Bend tunnel, Warrandyte © 2017 (picture section); Ballarat sludge drain © 2012 (picture section); sludge dam drawing and specification No.8 © 1907, Sludge Abatement Board.

Susan Lawrence: Cocks Eldorado dredge © 2012 (p. 59).

Public Records Office Victoria: Dunolly goldfield commons © Philip Chauncy 1862 (p. 109); Donald Fletcher's race network (application for a water right licence) © Henry Davidson 1883 (p. 121); John Pund & Co.'s application for water-right licence © Henry Davidson 1881 (p. 123); Petition to the Borough of Creswick from the Chinese miners at Black Lead © 1871 (p. 147); Friedrich Kassebaum's application for a water right licence © Henry Davidson 1881 (p. 170); Sludge Abatement Board letterhead © 1974 (p. 197); Map of Ballarat East sludge dam © Sludge Abatement Board, 1905 (p. 202).

State Library of Victoria: Map of Victorian Mining Districts © Victoria Department of Crown Lands and Survey 1866 (picture section); Sheepwash Creek near Bendigo © Francois Cogné 1863, pictures collection (p. 12); Horse Puddling Machine, Forest Creek © S.T. Gill 1855, pictures collection (p. 15); Sludge in Spring Creek, Daylesford © Richard Daintree and Antoine Fauchery 1859, pictures collection (p. 17); Black Hill, Ballarat © Richard Daintree 1994, pictures collection (p. 20); Ground sluicing © James Stirling c.1800–1900, pictures collection (p. 39); Box sluicing © Richard Daintree c.1861, pictures collection (p. 40); Chinese gold workings at Guildford along the Loddon © Richard Daintree, pictures collection (picture section); Hydraulic sluicing at Pioneer Claim open cut mine, Mitta Mitta © unknown, c.1860–1870, pictures collection (p. 45); Hydraulic sluicing at Oriental Claims, Omeo © Walter Hodgkinson c.1880–1900, pictures collection (p. 45); Garfield waterwheel, Chewton © Lorraine Photographers c.1937, pictures collection (p. 56); Flume on the Buckland © Sherwin Engraver 1866, pictures collection (p. 91); Bendigo sludge © John Henry Harvey c.1890–1938, pictures collection (p. 148); Sludge lake and mining operations at Long Gully, Bendigo © Charles Daniel Pratt c.1925–1940, pictures collection (p. 202).

Department of Environment, Land, Water and Planning (formerly Department of Environment and Primary Industries): LiDAR imagery of Three Mile Creek © 2010, accessed via La Trobe University (p. 37).

Parliament of Victoria: Sir John Bowser © c.1920 (p. 179).

State Library of New South Wales: The Precious Metal © J.A. Turner 1894, courtesy of Dixson Galleries, DG314 (picture section).

National Library of Australia: Stanley, Ovens © E. Hulme 1859, nla.obj-135946012 (picture section).

Eldorado Museum: Eldorado hydraulic pump sluicing © unknown (p. 59).

Burke Museum: Percy Pund and miners at Three Mile, Beechworth © unknown artist (picture section).

NOTES

Introduction

1 'The Brumadinho Tailings Dam Failure', World Information Service on Energy website, http://www.wise-uranium.org/mdafbr.html, accessed 18 March 2019.

2 Griffiths, T. 2016, *The Art of Time Travel: Historians and their Craft*, Black Inc., Melbourne, 5.

3 Phillips, G.N., Hughes, M.J., Arne, D.C., Bierlein, F.P., Carey, S.P., Jackson, T. and Willman, C.E. 2003, 'Gold: Historical Wealth, Future Potential', in W.D. Birch (ed.), *Geology of Victoria*, Geological Society of Australia Special Publication 23, Geological Society of Australia (Victoria Division), 377; 380; 414.

1 Sludge

1 *Report of the Board Appointed by His Excellency the Governor in Council to Inquire into the Sludge Question* 1887, Parliament of Victoria, Melbourne, 120–123.

2 *Mount Alexander Mail* 9 September 1870: 2c–d.

3 Public Records Office Victoria, VA 2889, Registrar-General's Department, VPRS 24/P0 Inquest Deposition Files, unit 212, item 1868/914 Male; unit 115, item 1862/803 Male; unit 77, item 1860/70 Male.

4 Bate, W. 2001, 'Gold: Social Energiser and Definer', *Victorian Historical Journal* 72(1–2): 7–27.

5 *Geelong Advertiser*, 19 April 1859: 2c.

6 Sludge Select Committee 1861 (1887), Minutes of Evidence, Parliament of Victoria, Melbourne, 16–17; Dunstan, D. 1994, *Better than Pomard! A History of Wine in Victoria*, Australian Scholarly Publishing, Melbourne, 101; *The Age*, 2 November 1861: 7f.

7 Fahey, C. 1982, 'The Ballerstedts and the Bendigo Quartz Reefs', *La Trobe Journal* 30: 29–33; Harvey, R.C. 1991, *Background to Beechworth From 1852*, Beechworth Progress Association, Beechworth, 7; Taylor, A. 1998, *A Forester's Log: The Story of John La Gerche and the Ballarat-Creswick*

State Forest 1882–1897, Melbourne University Press, Melbourne, 3.

8   Mayes, C. 1861, 'Essay on the Manufactures More Immediately Required
    for the Economical Development of the Resources of the Colony', in
    Victorian Government Prize Essays, Government Printer, Melbourne, 354.

9   Sludge Board 1887: xxiv.

10 J. Cameron, *Victoria Parliamentary Debates* Legislative Assembly
    27 October 1903, 928.

11 *Report of the Royal Commission Appointed to Enquire into the Best
    Method of Removing the Sludge from the Gold Fields* 1859, Parliament
    of Victoria, Melbourne, 26.

12 Sludge Royal Commission 1859: 3–5.

13 Sludge Select Committee 1861 (1887): 16–17.

14 Sludge Select Committee 1861 (1887): 3–4

15 Sludge Royal Commission 1859: 5.

16 Peterson, L. 1996, *Reading the Landscape: Documentation and Analysis
    of a Relict Feature of Land Degradation in the Bendigo District, Victoria,*
    Monash Publications in Geography and Environmental Science, Monash
    University, Melbourne, 44–45; 75; 76; 82.

17 Bate, W. 1978, *Lucky City. The First Generation at Ballarat: 1851–1901,*
    Melbourne University Press, Melbourne, 99–101; *Geelong Advertiser,* 27
    April 1859: 3e.

18 Brown, P.L. and Russell, G. 1941–1971, Clyde Company Papers V 1854–8,
    Oxford University Press, London, 164.

19 *Ballarat Star,* 25 October 1869: 6f.

20 Sludge Board 1887: 139.

21 *Sludge – Inverleigh and Shelford, Return to an Order of the Legislative
    Assembly,* 16 October 1872, Parliament of Victoria, Melbourne 771–774;
    Report of the Sludge Abatement Board for 1909 (1910), Plates O; S.

22 *The Argus,* 19 October 1934: 3g.

23 Reeves, J., Gell, P., Reichman, S., Trewarn, A. and Zawadski, A. 2015,
    'Industrial Past, Urban Future: Using Palaeo-studies to Determine the
    Industrial Legacy of the Barwon Estuary, Victoria, Australia', *Marine
    and Freshwater Research* 67(6): 837–849.

24 Davies, P., Lawrence, S. and Turnbull, J. 2015, 'Mercury Use and Loss
    From Gold Mining in Nineteenth-Century Victoria', *Proceedings of
    the Royal Society of Victoria* 127: 44–54; Environmental Protection
    Authority 2016, 'Mercury and Arsenic in Victorian Waters: A Legacy of
    Historical Gold Mining', in *Environmental Protection Authority Victoria
    Publications 25,* Carlton; Rae, I.D. 2001, 'Gold and Arsenic in Victoria's
    Mining History', *Victorian Historical Journal* 72(1–2): 159–172; Sultan,
    K. 2006, 'Distribution of Arsenic and Heavy Metals in Soils and Surface
    Waters in Central Victoria (Ballarat, Creswick and Maldon)', PhD thesis,
    School of Science and Engineering, University of Ballarat, Ballarat.

25 Public Records Office Victoria, Creswick Council Minute Books, 7 June 1859, VPRS 3730, Ballarat Archives Centre.

26 *Bendigo Advertiser*, 1 November 1860: 2b–c.

27 Sludge Board 1887: 51–60.

28 Sludge Board 1887: xx.

29 Sludge Board 1887: xii.

30 *Ballarat Star*, 17 November 1881: 2d.

31 Sludge Board 1887: 26–27.

32 Sludge Board 1887: 51–60; Woodland, J. 2001, *Sixteen Tons of Clunes Gold: A History of the Port Phillip and Colonial Gold Mining Company*, Clunes Museum, Clunes, Victoria, 68–69; *Argus*, 14 November 1881: 6f.

33 *Bairnsdale Advertiser and Tambo and Omeo Chronicle*, 28 March 1893: 2h.

34 *Bairnsdale Advertiser and Tambo and Omeo Chronicle*, 24 May 1904: 2d; 16 December 1909: 3b; 24 June 1911: 2f.

35 *Gippsland Times*, 1 July 1865: 2e; *Traralgon Record*, 18 February 1898: 2.

36 *Bairnsdale Advertiser and Tambo and Omeo Chronicle*, 14 August 1906: 2f, 16 October 1906: 4a; *Alexandra and Yea Standard and Yarck, Gobur, Thornton and Acheron Express*, 6 May 1910: 2g; *Seymour Express and Goulburn Valley Avenel, Graytown, Nagambie, Tallarook and Yea Advertiser*, 13 October 1917: 3b.

37 *Bendigo Independent*, 30 August 1909: 3a–e.

38 Davis, J., Rutherfurd, I. and Finlayson, B. 1997, 'Reservoir Sedimentation Data in South Eastern Australia', *Water* March/April: 11–15.

39 Davis, J. and Finlayson, B. 2000, *Sand Slugs and Stream Degradation: The Case of the Granite Creeks, North-East Victoria*, Cooperative Centre for Freshwater Ecology Technical Report 7/2000, 1; Abernathy, B., Markham, A., Prosser, I.P. and Wansbrough, T. 2004, 'A Sluggish Recovery: The Indelible Marks of Landuse Change in the Loddon River Catchment', in *Fourth Australian Stream Management Conference: Linking Rivers to Landscapes*, Launceston, Tasmania, 19–22.

40 Sludge Board 1887: 22.

41 See for e.g. *Mount Alexander Mail*, 6 January 1912: 2f.

42 *The Leader*, 11 February 1865: 17a; Sludge Board 1887: 127.

43 *Bendigo Advertiser*, 14 April 1869: 2e; Public Records Office Victoria, VA 2889, Registrar-General's Department, VPRS 24/P0 Inquest Deposition Files, unit 223, file 1869/393 Male.

44 *The Argus*, 26 March 1879: 7e; *The Age*, 28 March 1910: 6e.

45 *The Argus*, 12 October 1901: 15e.

46 Sludge Board 1887: xxviii.

47 *Ovens and Murray Advertiser*, 5 August 1876: 5b.

48 *Ovens and Murray Advertiser*, 21 December 1875: 2a–b.

49 *Ovens and Murray Advertiser*, 22 June 1878: 8a.

50 Tully, J. 2016, *Talbot Maps*, Weila Publishing, Dunolly, Victoria, 2; Sludge Board 1887: 63.

51 Sludge Board 1887: 143–159.

52 Sludge Board 1887: 14.

53 Sludge Board 1887: 159.

54 Sludge Board 1887: 155.

55 Sludge Board 1887: 1.

56 Sludge Board 1887: 145

57 Sludge Board 1887: xxvi.

58 Dredging and Sluicing Inquiry Board 1914, *Report upon Complaints of Injury by Dredging and Sluicing,* Parliament of Victoria, Melbourne, 10.

59 Dredging and Sluicing Inquiry Board 1914: 10.

60 Peterson, *Reading the Landscape*, 63–64.

61 Peterson, *Reading the Landscape*, 63–64.

62 A.A. Billson, *Victoria Parliamentary Debates*, Legislative Assembly, 9 August 1904, 715; James Campbell, *Victoria Parliamentary Debates*, Legislative Council, 17 December 1885, 2505; W.D. Smith, *Victoria Parliamentary Debates,* Legislative Assembly, 15 November 1881, 718.

63 *The Argus* 2 October 1875: 7d; 14 November 1881: 6f; Sludge Board 1887: 50–51

64 Dredging and Sluicing Inquiry Board 1914: 10–13; Davies et al., 2018, 'Environmental History of Bucket Dredging in Victoria', *Journal of Australasian Mining History* 16: 59–74.

65 *Ovens and Murray Advertiser*, 14 November 1903: 8g–h.

66 *Wangaratta Chronicle*, 6 February 1915: 3c.

67 T.R. Ashworth, *Victoria Parliamentary Debates*, Legislative Assembly, 28 October 1903, 970.

68 Dredging and Sluicing Inquiry Board 1914: 9.

69 *Ovens and Murray Advertiser*, 14 November 1903: 8g–h; Dredging and Sluicing Inquiry Board 1914: 8

70 *The Age*, 24 January 1934: 9d–e.

2 Mining

1 Lloyd, B. 2006, *Gold in the North-East: A History of Mining for Gold in the Old Beechworth Mining District of Victoria*, Histec Publications, Hampton East, Victoria, 156; *Ovens and Murray Advertiser*, 24 July 1915: 3c.

2 Lawrence, S. and Davies, 2015, 'Cornish Tin-Streamers and the Australian Gold Rush: Technology Transfer in Alluvial Mining', *Post-Medieval Archaeology* 49(1): 99–113.

3 Smyth, R.B. 1980, *The Gold Fields and Mineral Districts of Victoria*, facsimile of 1869 edition, Queensberry Hill Press, Melbourne, 127; Russell, G. 2009, *Water for Gold! The Fight to Quench Central Victoria's Goldfields*, Australian Scholarly Publishing, Melbourne, 242.

4   *The Argus*, 13 May 1853: 6g.
5   *Report from the Select Committee on Castlemaine and Sandhurst Water Supply* 1864–1865, Parliament of Victoria, Melbourne, 9–10.
6   *Goulburn Herald*, 7 May 1853: 3a.
7   *Report from the Select Committee of the Legislative Council on Gold Mining on Private Property* 1855–1856, Parliament of Victoria, Melbourne, 8.
8   Deason, D. 2005, *Welcome Stranger: The Amazing True Story of One Man's Legendary Search for Gold at all Costs,* Penguin/Viking, Melbourne, 35–38.
9   Payton, P. 1984, *The Cornish Miner in Australia (Cousin Jack Down Under)*, Dyllansow Truran, Cornwall, 38–39; 112; 117; 120; Croggan, J. 2002, 'Strangers in a Strange Land – Converging and Accommodating Celtic Identities in Ballarat 1851–1901', PhD thesis, University of Ballarat; Hopkins, R. 1988, *Where Now, Cousin Jack?* Bendigo Building Society, the City of Bendigo and the Australian Bicentennial Authority, Bendigo, Vic; Fahey, C. 2007, 'From St Just to St Just Point: Cornish Migration in Nineteenth-Century Victoria', *Cornish Studies*: 15; 121.
10   Blainey, G. 1963, *The Rush That Never Ended: A History of Australian Mining*, Melbourne University Press, Melbourne, 87–88; Birrell, R.W. 2005, 'The Development of Mining Technology in Australia 1801–1945', PhD thesis, University of Melbourne, Melbourne, 57–58; Yip, Y.H. 1969, *The Development of the Tin Mining Industry of Malaya*, University of Malaya Press, Kuala Lumpur, 67–89; McGowan, B. 1996, 'The Typology and Techniques of Alluvial Mining: 'The Example of the Shoalhaven and Mongarlowe Goldfields in Southern New South Wales', *Australasian Historical Archaeology* 14: 34–45; Rolls, E. 1992, *Sojourners: The Epic Story of China's Centuries-Old Relationship with Australia*, University of Queensland Press, Brisbane, 102; Serle, G. 1963, *The Golden Age: A History of the Colony of Victoria, 1851–1861*, Melbourne University Press, Melbourne, 320; Smith, L. 2003, 'Identifying Chinese Ethnicity Through Material Culture: Archaeological Excavations at Kiandra, New South Wales', *Australasian Historical Archaeology* 21: 18–29; Talbot, D. 2004, *The Buckland Valley Goldfield*, Specialty Press, Albury, NSW.
11   *Mineral Statistics of Victoria for the Year 1884*, 1885, 54–55.
12   *Mount Alexander Mail*, 2 October 1905: 2.
13   *Mining Surveyors' Reports*, July 1859: 8.
14   *Mineral Statistics of Victoria for the Year 1884*, 1885: 54–55.
15   *Report of the Board Appointed by His Excellency the Governor in Council to Inquire into the Sludge Question 1887*, Parliament of Victoria, Melbourne, 3; *Victorian Government Gazette* 19 August 1881: 2422; Public Records Office Victoria, VPRS 6784/P0006 Water Right Licence Files 1863–1973, Unit 3, File 626WR, Plan of 20a 3r 6p Applied for on Licence Under the Water Right Licence Regulations by John Pund & Co.

Application for Licence by John Pund & Co, 28 January 1881, Victorian Archives Centre, Melbourne.

16 Smyth, *The Gold Fields and Mineral Districts of Victoria*, 130.

17 *Creswick Advertiser*, 2 November 1860; Chin, I. and Scott, C. 2009, *Coronial Inquests and Magisterial Inquiries. Creswick Chinese (1856 to 1905)*, The Chinese Heritage Interest Network, Blackburn South, Victoria.

18 *Report of the Chief Inspector of Mines to the Honorable the Minister of Mines*, Parliament of Victoria, Melbourne, 1874–1883.

19 Scott, E. (ed) 1945, *Lord Robert Cecil's Gold Field's Diary*, 2nd edition, Melbourne University Press, Melbourne, 27.

20 *Report of the Royal Commission Appointed to Enquire into the Best Method of Removing the Sludge from the Gold Fields 1859*, Parliament of Victoria, Melbourne, 26.

21 *Bendigo Advertiser*, 26 June 1858: 2d.

22 Penney, J. 2001, 'The Creswick Mining Disaster of 1882: A Story of Loss, Mateship and Courage', *Victorian Historical Journal* 72(1–2): 187–203.

23 *Second Progress Report of the Royal Commission on Gold Mining*, 1891, Parliament of Victoria, Melbourne, 10.

24 Bower, C. 2013, *Water Races and Tin Mines of the Toora District*, Colleen Bower.

25 For e.g. Knighton, A.D. 1987, 'Tin Mining and Sediment Supply to the Ringarooma River, Tasmania 1875–1979', *Australian Geographical Studies* 25(1): 83–97.

26 Birch, W.D. and Henry, D.A. 2013, *Gemstones in Victoria*, Museum Victoria, Melbourne.

27 *Ballarat Star*, 21 December 1858: 2; Woodland, J. 2001, *Sixteen Tons of Clunes Gold: A History of the Port Phillip and Colonial Gold Mining Company*, Clunes Museum, Clunes, Victoria, 56.

28 Bland, R.H. 1890, *History of the Port Phillip and Colonial Gold Mining Company, in connection with the Clunes Mine*, F.W. Niven & Co, Ballarat, 11; *Mining Surveyors' Report*, March 1867: 11; Woodland, *Sixteen Tons of Clunes Gold*, 69.

29 Smyth, *The Gold Fields and Mineral Districts of Victoria*, 517.

30 Stoney, C. 1993, *The Howqua Hills Story*, Mansfield Historical Society, Mansfield, Victoria, 22; *Mansfield Guardian*, 21 October 1882.

31 Lawrence, S., Hoey, K. and Tucker, C. 2000, 'Archaeological Evidence of Ore Processing at the Howqua Hills Goldfield', *The Artefact* 23: 9–21.

32 Griffiths, T. 1992 *Secrets of the Forest: Discovering History in Melbourne's Ash Range*, Allen & Unwin, Sydney, 27; *Gold Fields Statistics, 1863* 1864, Parliament of Victoria, Melbourne, 8.

33 Russell, G., 2009 *Water for Gold! The Fight to Quench Central Victoria's Goldfields*, Australian Scholarly Publishing, Melbourne.

34   *Select Committee on Castlemaine and Sandhurst Water Supply,* 1864–1865, 11–17.

35   Davies, P. and Lawrence, S. 2013, 'The Garfield Water Wheel: Hydraulic Power on the Victorian Goldfields', *Australasian Historical Archaeology* 31: 25–32.

36   Smyth, *The Gold Fields and Mineral Districts of Victoria,* 28.

37   For e.g. *Dicker's Mining Record,* November 1862: 2; 80.

38   *Bendigo Advertiser,* 1 February 1890: 3.

39   Bower, *Water Races and Tin Mines of the Toora District,* 100.

40   Sellars, D.B. 1907, 'Dredge Mining and Hydraulic Sluicing', in *Annual Report of the Secretary for Mines and Water Supply for the Year 1906,* 94.

41   Supple, R. 1994, 'Cocks Eldorado Dredge', in *First Australasian Conference on Engineering Heritage 1994: Old Ways in a New Land; Preprints of Papers,* Institute of Engineers, Canberra, 158–159; Sheppard, D. 1982, *El Dorado of the Ovens Goldfields,* Research Publications, Blackburn, Victoria, 76–79.

42   Lloyd, B. 1982, *Gold at Harrietville,* Shoestring Press, Wangaratta, Victoria, 172–182.

43   McGeorge, J.H.W. 1964, *Dredging for Gold,* Barwon Heads, Victoria, 50–53.

44   Canavan, F. 1988, *The Deep Lead Gold Deposits of Victoria,* Bulletin 62, Geological Survey of Victoria, Melbourne, 36; Supple, 'Cocks Eldorado Dredge', 157.

45   Dredging and Sluicing Inquiry Board 1914, *Report upon Complaints of Injury by Dredging and Sluicing,* Parliament of Victoria, Melbourne, 10.

46   Sellars, D.B. 1913, 'Dredge Mining and Hydraulic Sluicing', in *Annual Report of the Secretary for Mines for the Year 1913,* 82.

47   McGeorge, *Dredging for Gold,* 73.

48   McGeorge, *Dredging for Gold,* 73.

49   *Ovens and Murray Advertiser,* 12 October 1901: 7a; *The Age,* 19 January 1906: 8b; *The Age,* 14 June 1906: 8c.

50   *The Age,* 24 January 1934: 9d–e.

51   *Report of the Royal Commission Appointed to Enquire into the Best Method of Removing the Sludge from the Gold Fields,* 1859, Parliament of Victoria, Melbourne, 36.

52   Sludge Royal Commission 1859, 49–50.

53   Davis, J. and Finlayson, B. 2000, *Sand Slugs and Stream Degradation: The Case of the Granite Creeks, North-East Victoria,* Cooperative Research Centre for Freshwater Ecology, Canberra, 1.

54   *Report from the Select Committee on Castlemaine and Sandhurst Water Supply* 1864–1865, Parliament of Victoria, Melbourne, 13, Q. 298.

55   Dingle, A.E. and Rasmussen, C. 1991, *Vital Connections: Melbourne and its Board of Works, 1891–1991,* McPhee Gribble, Melbourne, 29.

3 Water

1   *The Argus*, 13 May 1853: 6g.

2   *The Argus*, 13 May 1853: 6g; *Report from the Select Committee on Castlemaine and Sandhurst Water Supply*, 1864–1865, Parliament of Victoria, Melbourne, 15.

3   *The Argus*, 13 May 1853: 6g.

4   Alpha [Charles Grey Bird] 1915, *Reminiscences of the Goldfields of Victoria, New Zealand and New South Wales in the Fifties and Sixties (in Two Parts)*, Part 1, Gordon and Gotch, North Melbourne, 36.

5   *Select Committee on Castlemaine and Sandhurst Water Supply*, 15.

6   *Sydney Morning Herald*, 9 June 1853: 2c; 23 July 1853: 5b.

7   *Geelong Advertiser and Intelligencer*, 12 April 1853: 2c; *Supplement to the Victoria Government Gazette*, 8 April 1853: 483–5.

8   *The Argus*, 13 May 1853: 6g.

9   Smyth, R.B. 1980, *The Gold Fields and Mineral Districts of Victoria*, facsimile of 1869 edition, Queensberry Hill Press, Melbourne, 549.

10 *Dicker's Mining Record*, 24 December 1862: 18.

11 Nathan, E. 2007, *Lost Waters: A History of a Troubled Catchment*, Melbourne University Press, Melbourne, 10–13.

12 Russell, G. 2009, *Water for Gold! The Fight to Quench Central Victoria's Goldfields*, Australian Scholarly Publishing, Melbourne, 208–209.

13 Powell, J.M. 1989, *Watering the Garden State: Water, Land and Community in Victoria 1834–1988*, Allen & Unwin, Sydney, 6.

14 Gammage, B. 2011, *The Biggest Estate on Earth: How Aborigines Made Australia*, Allen & Unwin, Sydney, 103–107.

15 Russell, G. 2005, 'Liquid Gold: Bendigo's Water Supply', in M. Butcher and Y.M.J. Collins (eds), *Bendigo at Work: An Industrial City*, Holland House for the National Trust of Australia, Victoria, 115.

16 Russell, *Water for Gold*, 22; 37.

17 Russell, *Water for Gold*, 16. One of the worst outbreaks of disease was on the Buckland in 1854–1855, when more than a thousand miners died from typhoid (Serle, G. 1968, *The Golden Age: A History of the Colony of Victoria, 1851–1861*, Melbourne University Press, Melbourne, 80).

18 Lindsay, D. 1965, *The Leafy Tree*, F.W. Cheshire, Melbourne, 24.

19 *Ballarat Star*, 2 July 1862: 4a.

20 *Creswick Advertiser*, 5 December 1933.

21 Smith, N. 1971, *A History of Dams*, Peter Davies, London, 195–207.

22 *Illustrated Sydney News* 5.11.1853:3c.

23 *Illustrated Sydney News*, 5 November 1853: 3c; 3 June 1854: 7b.

24 *Ballarat Star*, 14 November 1858: 2g; *Creswick Advertiser*, 20 April 1860: 2; 27 April 1860: 3; Archer, W.H. 1868, *Indexes to Patents Registered in Victoria 1854–1866*, Victorian Government, Melbourne, 85; *Ballarat Star*, 4 July 1857: 2e.

25 *Report from the Select Committee on the Navigation of the Murray, &c.* 1858 Votes and Proceedings of the Legislative Assembly of New South Wales, volume 3, Sydney, 772.

26 Benjamin Franklin Eaton, Victorian Death Certificate, 11 December 1894, registration no. 12951.

27 *Dicker's Mining Record*, 24 January 1863: 37.

28 *Creswick Advertiser*, 3 June 1862: 2; 4 November 1859: 4.

29 Bannear, D. 1998, Historic Gold Mining Sites in Amherst Mining Division, Maryborough Mining District, Department of Natural Resources and Environment, Melbourne, 45–46; Flett, J. 1975, *Maryborough, Victoria: Goldfields History*, Poppet Head Press, Glen Waverley, Victoria, 68; Land Conservation Council Victoria 1980, *Report on the Ballarat Area*, 200–201.

30 Tracey, M.M. 1997, 'No Water – No Gold: Hydrological Technology in Nineteenth Century Gold Mining – An Archaeological Examination', in R. Kerr and M.M. Tracey (eds), *The Australian Historical Mining Association Conference Proceedings 1996*, Home Planet Design and Publishing, Canberra, 7; Smyth, *Gold Fields and Mineral Districts of Victoria*, 131.

31 Smyth, *The Gold Fields and Mineral Districts of Victoria*, 141.

32 *Report of the Royal Mining Commission Appointed to Enquire into the Conditions and Prospects of the Gold Fields of Victoria 1862–62*, Parliament of Victoria, Melbourne, evidence from J. Flinn, 486.

33 *Mining By-Laws of and for the State of Victoria 1916*, 21 (published in *Victoria Government Gazette*, 13 November 1916, 4333–4356).

34 *Report from the Select Committee of the Legislative Council on Gold*, 1856, D. – No. 18a, Parliament of Victoria, Melbourne, 8.

35 *Report on the Ovens Gold Fields Water Company's Bill 1859–60*, Parliament of Victoria, Melbourne, 10.

36 Smyth, *The Gold Fields and Mineral Districts of Victoria*, 405.

37 Sankey, R. H. 1871, *Report on the Coliban and Geelong Schemes of Water Supply*, Government of Victoria, Melbourne, 107. In later years a sluice-head was standardised as one cubic foot of water per second.

38 *Mount Alexander Mail*, 11 March 1859: 4c.

39 *Dicker's Mining Record*, 24 December 1862: 21; 24 December 1863: 262–263.

40 Sankey, *Coliban and Geelong Schemes of Water Supply*, 67–68. Performing routine maintenance on these mining dams could also be hazardous. Following heavy rain in November 1887 at Bendigo, William Rae went out to repair a breach in one of the dam banks serving his mine on the Victoria reef. The exertion brought on a pain in his side and a few days later he died of an internal obstruction (*Bendigo Advertiser*, 25 November 1887: 2f).

41 Macartney, J.N. 1871, *The Bendigo Goldfield Registry*, Charles F. Maxwell, Melbourne, 70–72; *Mineral Statistics of Victoria for the Year 1871*, 1872, 40.

42 *Victoria Government Gazette*, 19 August 1881: 2422; Sludge Board, *Report of the Board Appointed By His Excellency the Governor in Council to Inquire Into the Sludge Question*, 1887, Government of Victoria, Melbourne, 3.

43 Smyth, *Gold Fields and Mineral Districts of Victoria*, 548.

44 Public Records Office Victoria, VPRS 6784/P0006 Water Right Licence Files 1863–1973, Unit 3, File 626WR, Plan of 20a 3r 6p Applied for on Licence Under the Water Right Licence Regulations by John Pund & Co. Application for Licence by John Pund & Co, 28 January 1881, Victorian Archives Centre, Melbourne.

45 Lloyd, B. 2006, *Gold in the North-East: A History of Mining for Gold in the Old Beechworth Mining District Of Victoria*, Histec Publications, Hampton East, Victoria, 156.

46 *The Age*, 11 January 1864: 6b; Bannear, D. 1998, 'Historic Gold Mining Sites in Amherst', 46; 'Death of Mr. J.S. Stewart, M.L.A.', *The Age*, 13 November 1889: 5e; *Reports of the Mining Surveyors and Registrars. Quarter Ending 30th June 1871*, 33.

47 *Dicker's Mining Record* II, 1863: 177; Maddicks, H.T. 2016, *100 Years of Daylesford Gold Mining History*, revised edition, Daylesford and District Historical Society, Daylesford, 56; Davies, P., Lawrence, S. and Turnbull, J. 2016, 'The River Loddon and Tributaries Water Supply Company', *Journal of Australasian Mining History* 14: 21–36.

48 Alpha, *Reminiscences of the Goldfields*, 35–6; *The McIvor Times and Rodney Advertiser*, 19 October 1882: 3b–c; Randell, J.O. 1985, *McIvor: A History of the Shire and the Township of Heathcote*, published by the author, 111; Bannear, D. 1993, 'Historic Mining States in the Heathcote (Waranga South) Mining Division, North Central Goldfields Project', Department of Conservation and Natural Resources North West Area, Victoria, 13–14.

49 Wynn, G. 1979, 'Life on the Goldfields of Victoria: Fifteen Letters', *Journal of the Royal Australian Historical Society* 64 (4): 264.

50 Wynn, 'Life on the Goldfields of Victoria', 264–265.

51 Wright, C. 2013, *The Forgotten Rebels of Eureka*, Text, Melbourne, pp. 62–63.

52 McMahon, H.D. and Wild, C.G. 2008, *American Fever Australian Gold: American and Canadian Involvement in Australia's Gold Rush*, H.D. McMahon and C.G. Wild, Middle Park, Qld; *The Argus*, 8 January 1853: 3e.

53 Scholefield, G.H. (editor) 1940, 'Robertson, James William', in *A Dictionary of New Zealand Biography* Volume 2, Department of Internal Affairs, Wellington, 246–247; De La Mare 2001, 'Lake Shipping and Daniel McBride', *The Queenstown Courier*, 67: 9–10.

54 *Creswick Advertiser*, 29 June 1860.

55 Mayes, C.H. 1861, 'Essay on The Manufactures More Immediately Required for the Economical Development of the Resources of the Colony', in *The Victorian Government Prize Essays*. Government Printer, Melbourne, 271–272.

56 Wynn, 'Life on the Goldfields of Victoria', 265.

57 Miller, F.W.G. 1966, *Golden Days of Lake County*, fourth edition, Whitcombe and Tombs, Christchurch, pp.123–124; Scholefield, Robertson, James William, 246–247.

58 *Annual Report of the Secretary for Mines for the Year 1911 (1912)*, 65.

59 Davies et al., 'The River Loddon and Tributaries Water Supply Company': 21–36; A.H. Bradfield, Work Diary of Arthur Bradfield for 1937. Manuscript in possession of Jeannie Lister, Castlemaine, Victoria.

60 Davies, P. and Lawrence, S. 2014, 'Bitumen Paper Pipes and Technology Transfer on the Victorian Goldfields', *Journal of Australasian Mining History* 12: 45–58.

61 *Creswick Advertiser*, 3 June 1862; 12 September 1862.

62 St. George's Sluicing Company, 'Memorial', 18 September 1862, Box 30, E. J. Semmens Collection, University of Melbourne Archives; *Creswick Advertiser*, 16 September 1862.

63 *Ararat and Pleasant Creek Advertiser*, 27 January 1863: 2–3.

64 *Creswick Advertiser*, 7 October 1862: 2d; *Ballarat Star*, 5 September 1864: 3b.

65 *Creswick Advertiser*, 24 October 1862: 2d.

66 Public Records Office Victoria, VPRS 398, *Correspondence of the Mining Registrar, Fryerstown*, Consignment P0000, File No. 4, Field Book 2a, October 1869, Victorian Archives Centre, Melbourne.

67 Brown, G.O. *Reminiscences of Fryerstown*, Castlemaine, Victoria, 1983, 212.

68 *Illustrated Australian News*, 13 August 1870: 142a; Houghton, M. 2009, *Warrandyte: The Community With a Heart of Gold*, Penfolk Publishing, Blackburn, Victoria, 214.

69 Letter from J. Flinn, Esq., Stanley, Ovens, 1 November 1862, paragraph 7, reproduced in *Report of the Royal Mining Commission 1862–1863*, 486; *Report from the Select Committee*, 1859–1860, 16, question 210.

70 Lloyd (*Gold in the North-East*, pp.21–34) discusses John Wallace's career in detail; see also Philipp, J. 1987, *Bethanga: A Poorman's Diggings: Mining and Community at Bethanga, Victoria, 1875–1912*, Hyland, Melbourne.

71 Shennan, R. 2004, *A Biographical Dictionary of the Pioneers of the Ovens and Townsmen of Beechworth*, M.R. Shennan, Noble Park, Victoria, 23.

72 *Report of the Royal Mining Commission 1862–1863*: 345–346.

73 Lloyd, *Gold in the North-East*, 42; Woods, C. 1985, *Beechworth: A Titan's Field*, Hargreen Publishing Company, Melbourne, 99.

74 *Mineral Statistics of Victoria for Year 1884*, 1885: 54–55.

75 *Ovens and Murray Advertiser* 30 September 1899: 7c.

76 Lloyd, *Gold in the North-East*, 21–22.
77 *Dicker's Mining Record,* 11 May 1868: 178; Lloyd, *Gold in the North-East* 2006, 27; *Report from the Select Committee of the Legislative Assembly, Upon the Ovens Gold Fields Water Company's Bill; Together with the Proceedings of the Committee and Minutes of Evidence,* 1859–1860, Government of Victoria, Melbourne, 10.
78 *Report of the Royal Mining Commission* 1862–1863: 345.
79 *Mineral Statistics of Victoria for the Year 1884,* 1885: 54; 'Graves in the Blackwood Cemetary', *Blackwood News* April/May 2010: 8.
80 *Mineral Statistics of Victoria for the Year 1884,* 1885: 54.
81 Kyi, A. 2004, 'Unravelling the Mystery of the Woah Hawp Canton Quartz Mining Company, Ballarat', *Journal of Australian Colonial History* 6: 59–78.
82 Chin, I. and Scott, C. 2009, *Coronial Inquests and Magisterial Inquiries: Creswick Chinese (1856 to 1905),* The Chinese Heritage Interest Network, Blackburn South, Victoria, 161.
83 Chin, M., Chin, I. and Scott, C. 2009, *Chinese in the Creswick Cemetery: Headstones & Inscriptions, A Cultural Interpretation,* The Chinese Heritage Interest Network, Blackburn South, Victoria, 104–111.
84 Sludge Board 1887: 21–23.
85 Sludge Board 1887: 123.
86 Goldsmith, L. 2000, *This Golden Life,* Likely Prospects, Daylesford, Victoria, 3–4.
87 Diary of William Crawford Walker, 1852–1863, State Library Victoria, MS 11485.
88 *Victorian Government Gazette,* 5 June 1863: 1253.
89 Smyth, *Gold Fields and Mineral Districts of Victoria,* 405.
90 Nathan, *Lost Waters,* 65–67.
91 *Geelong Advertiser,* 24 January 1872: 3.
92 Royal Commission 1885, *Royal Commission on Water Supply. Further Progress Report,* Parliament of Victoria, Melbourne, 203.
93 Davies, P. and Lawrence, S. 2018, 'Pioneers of Goldfields Water Management: The Lal Lal Waterworks Association', *Australasian Historical Archaeology* 36: 59–68.

4 Fist Fights and Water Rights

1 *Ovens and Murray Advertiser,* 29 May 1866: 3a–b.
2 Hilderbrand, J. 2012, *The Baarmutha Story,* second edition, PB Publishing, Gisborne, Victoria, 229–230.
3 *Ovens and Murray Advertiser,* 31 July 1872: 2d; 8 October 1872: 2g.
4 *Creswick Advertiser,* 27 July 1860: 26; 10 August 1860: 4.
5 *Ovens and Murray Advertiser,* 12 October 1865: 2g.
6 *Ovens and Murray Advertiser,* 30 November 1865: 2b; 5 December 1865: 2e.

7   Smyth, R.B. 1980, *The Gold Fields and Mineral Districts of Victoria*, facsimile of 1869 edition, Queensberry Hill Press, Melbourne, 398.

8   Harris, E. 2006, *Development and Damage: Water and Landscape Evolution in Victoria*, Australia, Landscape Research 31 (2): 169–181; Stoeckel, K., Webb, R., Woodward, L. and Hankinson, A. 2012, *Australian Water Law*, Thomson Reuters, Sydney, 17.

9   Harris, *Water and Landscape Evolution in Victoria*, 171–172.

10   Attwood, B. 2009, *Possession: Batman's Treaty and the Matter of History*, The Miegunyah Press, Melbourne, 40–88.

11   Davies, P., Lawrence, S. and Twigg, K. 2018, 'Grazing Was Not Mining: Managing Victoria's Goldfields Commons', *Geographical Research* 56 (3): 256–269; Davies, P., Twigg, K. and Lawrence, S. 2018, 'Common to All Miners: The Inglewood Gold Field Common', *Provenance: The Journal of Public Record Office Victoria* 2018, https://prov.vic.gov.au/explore-collection/provenance-journal/provenance-2018/common-all-miners.

12   Parliament of Victoria 1860–1861, *Crown Lands Sales Act. Proclamations Respecting Commons Under the Said* Act. 6 February 1861.

13   Cahir, F. 2012, *Black Gold: Aboriginal People on the Goldfields of Victoria, 1850–1870*, ANU EPress, Canberra.

14   Clarke, G. 1994, *Irish Fortunes: Clarke and Russell Families in Creswick*, K M & G Clarke, Waramanga, ACT, 27.

15   *Ballarat Star*, 2 May 1861 (Supplement): 1c.

16   *Ballarat Star*, 2 May 1861 (Supplement): 1c; *Ballarat Star*, 10 August 1858: 2g; 3 February 1859: 2f; *Creswick Advertiser*, 20 April 1860.

17   *Ballarat Star*, 6 June 1857: 2f.

18   *Ballarat Star*, 22 July 1857: 3d.

19   *Report of the Commission Appointed to Enquire into the Conditions of the Gold Fields of Victoria*, 1854–1855, Parliament of Victoria, Melbourne, 237.

20   Birrell, R. 1998, *Staking a Claim: Gold and the Development of Victorian Mining Law*, Melbourne University Press, Melbourne, 81–82.

21   Chin, M., Chin, I. and Scott, C. 2009, *Chinese in the Creswick Cemetery: Headstones & Inscriptions, A Cultural Interpretation*, The Chinese Heritage Interest Network, Blackburn South, Victoria, 98–130.

22   Talbott, D. 2004, The Buckland Valley Goldfield, Specialty Press, Albury, 87–107.

23   Trollope, A. 1968 [1873], *Australia and New Zealand*, Volume 1, Dawsons of Pall Mall, London, 82.

24   Hamilton, J.P. 2015, *Adjudication of the Goldfields in New South Wales and Victoria in the 19th Century*, Federation Press, Sydney, 207–211.

25   Blackstone, W. 1979, *Commentaries on the Laws of England*, facsimile of 1766 edition, vol. 2, University of Chicago Press, Chicago, 18.

26   Getzler, J. 2004, *A History of Water Rights at Common Law*, Oxford University Press, Oxford.

27 Cathcart, M. 2009, *The Water Dreamers: The Remarkable History of Our Dry Continent*, Text, Melbourne, 259.

28 Getzler, *History of Water Rights*, 204.

29 Blackstone, *Commentaries*, 403.

30 Kanazawa, M. 2015, *Golden Rules: The Origins of California Water Law in the Gold Rush*, The University of Chicago Press, Chicago; Schorr, D. 2012, *The Colorado Doctrine: Water Rights, Corporations, and Distributive Justice on the American Frontier*, Yale University Press, New Haven; Worster, D. 1985, *Rivers of Empire: Water, Aridity and the Growth of the American West*, Pantheon Books, New York.

31 *Victoria Government Gazette [Supplement]*, 8 April 1853: 483.

32 *Report of the Board Appointed to Inquire into and Report on Applications for Water Rights at Beechworth 1867*, Parliament of Victoria, Melbourne, 17.

33 Potts, E. D. and Potts, A. 1974, *Young America and Australian Gold: Americans and the Gold Rush of the 1850s*, University of Queensland Press, Brisbane, 53; *Further Papers Relative to the Recent Discovery of Gold in Australia, (In continuation of Papers presented 28th February 1853)*, Parliament of Great Britain, London, 187–188.

34 *Further Papers Relative to the Recent Discovery of Gold in Australia*, 180; 187–188.

35 Woods, C. 1985, *Beechworth: A Titan's Field*, Hargreen Publishing Company, Melbourne, 55; *Beechworth Water Rights: Report of the Commission Appointed to Enquire into the Subject of Water Rights in the Beechworth District*, 1860–1861, Parliament of Victoria, Melbourne, 2.

36 *Report of the Commission Appointed to Enquire into the Conditions of the Gold Fields of Victoria*, 1854–1855 Parliament of Victoria, Melbourne, xxxix.

37 Birrell, *Staking a Claim*, 35.

38 Woods, *Beechworth*, 54.

39 *Victoria Government Gazette*, 4 October 1855: 2500.

40 *Beechworth Water Rights*, 1860–1861: 2.

41 *Goldfields Amendment Act 1857*, 21 Vict. 32; s.3; s.11; s.76.

42 Barnard, C. 2012, *A Nineteenth Century Village-Yackandandah*, Blake Printing, Wodonga, Victoria, 2217.

43 *Ovens and Murray Advertiser*, 19 August 1858: 2e.

44 G.M.R. Rathbone 1969, Cope, Thomas Spencer (1821–1891), in D. Pike (ed), *Australian Dictionary of Biography*, vol. 3, Melbourne University Press, Melbourne, 457–459.

45 Royal Commission of Enquiry 1862–1863, *Report of the Royal Mining Commission Appointed to Enquire into the Conditions and Prospects of the Gold Fields of Victoria*, Parliament of Victoria, Melbourne, 370–371.

46 *Ovens and Murray Advertiser*, 27 August 1858: 3a–e.

47   Gold Fields Royal Commission of Enquiry 1862–1863: 8.

48   *Report on Applications for Water Rights at Beechworth* 1867: 5.

49   Public Records Office Victoria, VPRS 6784/P0006 Water Right Licence Files 1863–1973, Unit 3, File 626WR, Plan of 20a 3r 6p Applied for on License Under the Water Right License Regulations by John Pund & Co. Application for License by John Pund & Co, 28 January 1881, Victorian Archives Centre, Melbourne.

50   Royal Commission of Enquiry 1862–1863: 346.

51   Royal Commission of Enquiry 1862–1863: 358.

52   *Report from the Select Committee of the Legislative Assembly, Upon the Ovens Gold Fields Water Company's Bill, Together with the Proceedings of the Committee and Minutes of* Evidence 1859–1860, Parliament of Victoria, Melbourne, 24.

53   *Select Committee on the Ovens Gold Fields Water Company's Bill,* 1859–1860, 8–9.

54   For e.g. Public Records Office Victoria VA 2719 Department of Mines, VPRS 6784/P3 Water Right Licence Files 1863–1973, Unit 1, Water-Right Licence No.435, for Hambleton, Shandy & Party; Water-Right Licence No.437, for Thomas Little

55   Diary of William Crawford Walker, manuscript, State Library of Victoria.

56   *Acts, Orders in Council, Notices and Mining Board By-Laws, relating to the Gold Fields* 1874, Government Printer, Melbourne, 277.

57   Royal Commission of Enquiry 1862–1863: 348.

58   *Beechworth Water Rights* 1860–1861: 3. The connection between groundwater and surface water continues to be a source of dispute in many places today.

59   Webb, G.H.F. 1877, *The Victorian Law Reports,* Charles F. Maxwell, Melbourne, 1–17.

60   *Ovens and Murray Advertiser,* 13 October 1877: 2a; *Report from the Select Committee of the Legislative Assembly upon the Beechworth Waterworks Act 1860 Amendment Bill* 1877, Parliament of Victoria, Melbourne, 2.

61   Birrell, *Staking a Claim,* 103–104.

62   Woodland, J. 2001, *Sixteen Tons of Clunes Gold: A History of the Port Phillip and Colonial Gold Mining Company,* Clunes Museum, Clunes, Victoria, 29–32.

63   *Mining Statute* 1865 section 23.

64   *Beechworth Water Rights,* 1860–1861: 2.

65   *An Act to amend the Law relating to Leases of Auriferous Lands and for other purposes* 1862, section xi.

66   Smyth, *Gold Fields and Mineral Districts of Victoria,* 399.

67   *Acts, Orders in Council, Notices and Mining Board By-Laws, relating to the Gold Fields* 1874, Government Printer, Melbourne, 275.

68 VA 2719 Department of Mines, VPRS 6784/P3 Water Right Licence Files 1863–1973, Unit 1.

69 VA 2719 Department of Mines, VPRS 6784/P3 Water Right Licence Files 1863–1973, Unit 1.

70 *Dicker's Mining Record August 1863*: 191.

71 *Acts, Orders in Council 1874*: 434–435.

72 *Dicker's Mining Record August 1863*: 191.

73 *Acts, Orders in Council 1874*: 438–439.

74 *Acts, Orders in Council 1874*: 440.

75 Davies, P., Lawrence, S. and Turnbull, Jodi 2015, 'Historical Maps, Geographic Information Systems (GIS) and Complex Mining Landscapes on the Victorian Goldfields', *Provenance: The Journal of Public Record Office Victoria* 14.

76 Powell, J.M. 1970, *The Public Lands of Australia Felix: Settlement and Land Appraisal in Victoria 1834–1891*, Oxford University Press, Melbourne, 89–142; Wright, R. 1989, *'The Bureaucrats' Domain: Space and the Public Interest in Victoria, 1836–84*, Oxford University Press, Melbourne, 102–103.

77 *Victoria Government Gazette*, 16 April 1866: 818.

78 VA 2719 Department of Mines, VPRS 6784/P3 Water Right Licence Files 1863–1973, Unit 2. Water licences for mining purposes have also been permitted under the *Victorian Water Act* 1989.

79 Smyth, *Gold Fields and Mineral Districts of Victoria*, 405.

80 Curiously, water right licences issued under the Victorian *Water Act* 1989 also have this fifteen-year limit.

81 *Races, Dams, and Reservoirs. Gold Fields Act. Order in Council. Regulating Licences Authorising Persons to Cut, Construct, and Use Races, Dams, and Reservoirs 1862–1863*, Parliament of Victoria, Melbourne, Schedule F.

82 Worster, *Rivers of Empire*, 88.

83 Kanazawa, M. 2015, *Golden Rules: The Origins of California Water Law in the Gold Rush*, The University of Chicago Press, Chicago, 218.

84 Smyth, *Goldfields and Mineral Districts of Victoria*, 547–549.

85 Musgrave, W. 2008, 'Historical Development of Water Resources in Australia: Irrigation Policy in the Murray–Darling Basin', in L. Crase (ed), *Water Policy in Australia: The Impact of Change and Uncertainty*, Resources for the Future, Washington D.C., 30–31; Powell, J.M. 1989, *Watering the Garden State: Water, Land and Community in Victoria 1834–1988*, Allen & Unwin, Sydney, 113.

5 The Sludge Question

1 *Report of the Board Appointed by His Excellency the Governor in Council to Inquire into the Sludge Question* 1887, Parliament of Victoria, Melbourne, 152.

2   Shennan, R. 1990, *A Biographical Dictionary of the Pioneers of the Ovens and Townsmen of Beechworth*, M. Rosalyn Shennan, Noble Park, Victoria, 19–20.

3   Sludge Board 1887: 145.

4   Sludge Board 1887: vii.

5   Getzler, J. 2014, *A History of Water Rights at Common Law*. Oxford University Press, Oxford, 245–247.

6   Armstrong, H.J. 1901, *A Treatise on the Law of Gold Mining in Australia and New Zealand*, second edition, Melbourne, 224.

7   Acts, Orders in Council, Notices, and Mining Board By-Laws Relating to the Gold-Fields 1874, Minister of Mines, Melbourne, 250.

8   Acts, Orders in Council, 278; 364; 398; *Victoria Government Gazette* 111, 24 August 1858, 1611.

9   Bate W. 1988, *Victorian Gold Rushes*, McPhee Gribble/Penguin Books, Melbourne, 54; Blainey, G. 1963, *The Rush That Never Ended: A History of Australian Mining*, Melbourne University Press, Melbourne, p. 60; Serle, G. 1971, *The Rush to Be Rich: A History of the Colony of Victoria*, 1883–1889, Melbourne University Press, Melbourne, 63. In 1871 one-third of the colony's population lived in the goldfields region which had a total population of 270,428 – significantly more than Melbourne's 191,000 people.

10 Serle, *The Rush to be Rich*, 51; 80; Bate, *Victorian Gold Rushes*, 17–19.

11 Milner, P. 1986, 'Gold Mining and the Development of Engineering Firms in Victoria', *Royal Historical Society of Victoria Journal* 57(4): 11–22.

12 Historian Marjorie Theobald has written a detailed account of this episode, arguing for its significance as early evidence of attempts to protect the environment and the wider interests of the community. Theobald, M., 'Bull Versus the Puddling Machines: Castlemaine's First Environmental Cause', *Central Victorian Ecology*, 10 March 2013, http://centralvicecology.wordpress.com/2013/03/10/commissioner-bull-versus-the-puddling-machines-castlemaines-first-environmental-cause-marjorie-theobald/, accessed 8 April 2019.

13 Wright, C. 2013, *The Forgotten Rebels of Eureka*, Text, Melbourne, 2.

14 Sharwood, R.L. 1986, 'The Local Courts on Victoria's Gold Fields, 1855–1857', *Melbourne University Law Review* 15: 508–532.

15 Acts, Orders in Council, 226.

16 *The Ballarat Star*, 23 December 1857:2 g reported four such cases under the heading 'Puddlers Beware'.

17 *Ballarat Star*, 11 June 1858: 2b–c.

18 Hansen, Partnership Pty in association with Wendy Jacobs, Naga Services and Jan Penney 2003, Ballarat Heritage Study (Stage 2): Heritage Precincts, City of Ballarat, Bllarat, 170–184.

19 Public Records Office Victoria, VPRS 3730/P01, Council Minute Books, Creswick Council 1859.

20 According to historian Elizabeth Denny, the Creswick Chinese community was assertive and organised in its efforts to be recognised as part of local civic culture. Battles about sludge and water supply feature prominently in their campaign. Denny, E. 2012, 'Mud, Sludge and Town Water', *Provenance: The Journal of Public Record Office Victoria* no. 11.

21 Jean, M., Moloney, D. and the Castlemaine Historical Society 2015, Forest Creek to Forest Street heritage assessment report, Mount Alexander Shire Council, Castlemaine, 63–72.

22 Peterson, L. 1996, *Reading the Landscape: Documentation and Analysis of a Relict Feature of Land Degradation in the Bendigo District*, Victoria, Monash Publications in Geography and Environmental Science, Monash University, Melbourne, 63–64.

23 Legislative Assembly 1859–60, Sludge at Epsom, Parliament of Victoria, Melbourne.

24 Sludge at Epsom 1859–60, 7.

25 *The Argus*, 15 August 1873: 6f.

26 Cusack, F. 1973, *Bendigo: A History*, Heinemann, Melbourne, 118.

27 *Report of the Royal Commission Appointed to Enquire into the Best Method of Removing the Sludge from the Gold Fields 1859*, Parliament of Victoria, Melbourne, 7.

28 Cusack, Bendigo, 116–119.

29 Quaife, G.R. 1976, Sullivan, James Forester, in B. Nairn (ed), *Australian Dictionary of Biography* vol. 6, Melbourne University Press, Melbourne, 218–9.

30 *Bendigo Advertiser*, 12 May 1858: 2f; 13 May 1858: 2e.

31 Sludge Royal Commission 1859: 26.

32 Sludge Royal Commission 1859: 48.

33 Sludge Royal Commission 1859: 5.

34 Cusack, *Bendigo*, 119.

35 Wohl, E. 2004, *Disconnected Rivers: Linking Rivers to Landscapes*, Yale University Press, Yale, 185–196; Purseglove, J. 2015, *Taming the Flood: Rivers, Wetlands and the Centuries Old Battle Against Flooding*, William Collins Books, London, 331.

36 Rutherfurd, I. 2018, Options for the Management of Forest Creek through Castlemaine, report to the Castlemaine Landcare Group.

37 Vendargon, A. and Tate, B. 2012, Creswick Flood Mitigation and Urban Drainage Plan Final Report, Water Technology: Water, Coastal and Environmental Consultants, for North Central Catchment Management Authority and Hepburn Shire Council.

38 Parliament of Victoria 1862, *Auriferous Mining Leases Act. Order in Council. Regulations Under Act No. 148 Relating to Leases of Auriferous Lands and for other Purposes*, Schedule G.

39 Veatch, A.C. 1911, *Mining Laws of Australia and New Zealand*, Government Printing Office, Washington, 97.

40 *The Argus*, 29 December 1868: 5e; *Bacchus Marsh Express*, 10 August 1872: 4c; *Argus*, 17 October 1872: 6c

41 *Ballarat Star*, 25 October 1869: 2f.

42 *Ballarat Star*, 17 December 1868: 2c.

43 *The Argus*, 26 November 1869: 6f; Webb, G.H.F. (ed) 1874, *The Victorian Reports. Supreme Court of Victoria, at Law*. Sands & McDougall, Melbourne, 25–26.

44 *Ovens and Murray Advertiser*, 13 September 1870: 2f.

45 *Ovens and Murray Advertiser*, 2 July 1870: 2b; 2f; 10 August 1871: 3a; 21 December 1875: 2a; 30 June 1877: 4b.

46 *The Argus*, 14 June 1870: 5c; 7 September 1875: 5b; 18 November 1876: 7a; *Ovens and Murray Advertiser*, 31 January 1874: 1g.

47 *Ovens and Murray Advertiser*, 5 August 1876: 5b.

48 *Ovens and Murray Advertiser*, 30 June 1877: 4b; 3 December 1878: 2e; 19 April 1879: 8a.

49 Brown, P.L. 1967, Russell, George (1812–1888), in *Australian Dictionary of Biography* vol. 2, Melbourne University Press, Melbourne, 408–409.

50 Morres, H. 1872, *Sludge – Inverleigh and Shelford*, Parliament of Victoria, Melbourne.

51 Morres, *Sludge – Inverleigh and Shelford*, 2.

52 *The Argus*, 28 November 1872: 6b.

53 Veatch 1911, *Mining Laws of Australia and New Zealand*, 97.

54 For e.g. The Bonshaw Freehold Gold-Mining Company Registered v. The Prince of Wales Company Registered, in *Wyatt, Webb and A'Beckett's Reports* 1867, 126–133.

55 Atkins, J. 1871, *Notes on the Mining Statute, 1865*, printed for the author by W. S. Westneat, Prahran, Victoria, 13.

56 *The Argus*, 2 October 1875: 7d.

57 Canavan, F. 1988, *Deep Lead Gold Deposits of Victoria*, Geological Survey of Victoria, Bulletin 62, Melbourne, 34; McGeorge, J.H.W. 1966, *Buried Rivers of Gold*, the author, Melbourne, 35.

58 *The Age*, 12 Nov 1881: 6a.

59 *The Argus*, 1 November; 11 November; 20 December 1881; *The Age*, 12 November 1881.

60 Hansard vol. 37, 16 November 1881, 763.

61 Hansard vol. 38, 17 November 1881, 774–782.

62 Hansard vol. 38, 24 November 1881, 868.

63 Hansard vol. 38, 24 November 1881, 868.

64 Hansard vol. 38, 24 November 1881, 872.

65 *The Lands Compensation Statute* 1869, 33 Vict. 344, s. 64.

66 Powell, J. 1989, *Watering the Garden State: Water, land and community in Victoria 1834–1988*, Allen & Unwin, Sydney, p.113; Wright, R. 1989, *The Bureaucrats' Domain: Space and the Public Interest in Victoria, 1836–1884*, Oxford University Press, Melbourne, 225; Clark, S.D. 1971, 'Australian Water Law: An Historical and Analytical Background', PhD thesis, University of Melbourne, 162–163.

67 Hansard, vol. 38, 29 November 1881, 911.

68 *Mineral Statistics of Victoria for the Year 1881*, 19.

69 Hearn, T.J. 1982, 'Riparian Rights and Sludge Channels: A Water Use Conflict in New Zealand, 1869–1921', *New Zealand Geographer* 38: 51.

70 Hansard vol. 38, 24 November 1881, 777.

71 Hansard vol. 38, 24 November 1881, 781; 775; 883; 783.

72 Hansard vol. 38, 24 November 1881 p.872; 6 December 1881, 1010.

73 For e.g. *The Argus*, 19 February 1885: 6h; 24 October 1885: 13c; 30 December 1885: 9a.

74 *Victoria Government Gazette* 85, 6 August 1886, 2246.

75 Sludge Board 1887.

76 Sludge Board 1887: viii.

77 Sludge Board 1887: xxiii–xxviii.

78 Sludge Board 1887: xvii–xviii.

79 Sludge Board 1887: xxiii–xiv.

80 Sludge Board 1887: xxv.

81 Sludge Board 1887: 3; 8–11.

82 Sludge Board 1887: xxiv, xxv.

83 Sludge Board 1887: 3–14.

84 Davies, P., Lawrence, S. and Turnbull, J. 2016, 'The River Loddon and Tributaries Water Supply Company', *Journal of Australasian Mining History* 14: 21–36.

85 Davies, P. and Lawrence, S. 2013, 'The Garfield Water Wheel: Hydraulic Power on the Victorian Goldfields', *Australasian Historical Archaeology* 31: 25–32.

86 Sludge Board 1887: xxi, 73–85.

87 Sludge Board 1887: xxi.

88 Greenland, P. 2001, *Hydraulic Mining in California: A Tarnished Legacy*, Arthur H. Clark Company, Spokane, p.126; 156.

89 Sludge Board 1887: xiii.

90 Sludge Board 1887: xiii; Greenland, *Hydraulic Mining in California*, 255.

91 Sludge Board 1887: viii.

92 Sludge Select Committee 1861 (1887), Minutes of Evidence, Parliament of Victoria, Melbourne.

93 Sludge Board 1887: xi–xii.

94 Sludge Board 1887: ix.

95 Sludge Board 1887: ix.
96 Sludge Board 1887, Appendix A, Copy of Petition by Farmers and Others in Carisbrook District to the Honorable the Minister of Lands, xx.
97 Sludge Board 1887: xi.
98 Sludge Board 1887: viii.
99 Sludge Board 1887: xiv–xv.
100 Sludge Board 1887: ix–xiv.
101 Sludge Board 1887: xxviii–xxxi.
102 *The Argus*, 8 June 1892: 4h–5a; *Ovens and Murray Advertiser*, 26 November 1887: 15d.
103 *Progress Report of the Royal Commission on Gold Mining 1891*, Parliament of Victoria, Melbourne, 5–6.
104 Fahey, C. 1986, The Berry Deep Leads: An Historical Assessment, report submitted to Department of Conservation, Forests and Lands, Melbourne, 35.
105 Woodland, J. 2001, *Sixteen Tons of Clunes Gold: A History of the Port Phillip and Colonial Gold Mining Company*, Clunes Museum, Clunes, Victoria, 88–90.
106 Woods, C. 1985, *Beechworth: A Titan's Field*, Hargreen Publishing Company, Melbourne, 159.

6 Turning the Tide

1 *Ovens and Murray Advertiser*, 16 June 1906: 9a–d.
2 Vines, M. 1979, Bowser, Sir John (1856–1936), in B. Nairn and G. Serle (eds), *Australian Dictionary of Biography*, vol. 7, Melbourne University Press, Melbourne, 365–366.
3 Travis, J. *Annual Report of the Secretary for Mines and Water Supply During the Year 1899*, 1900, Parliament of Victoria, Melbourne, 6.
4 Dredging and Sluicing Inquiry Board 1914, *Report Upon Complaints of Injury by Dredging and Sluicing*, Parliament of Victoria, Melbourne, p.18; 'Desolating Dredges' headline used repeatedly by *The Age* in reporting, e.g. 18 June 1912: 10e.
5 *Ovens and Murray Advertiser*, 14 November 1903: 8g.
6 *Mount Alexander Mail*, 4 August 1905, p.2e; *Age* 18 June 1906, 6c.
7 Dredging and Sluicing Inquiry Board 1914: 13.
8 Dredging and Sluicing Inquiry Board 1914: 11.
9 *The Argus*, 3 August 1901: 14h; *Ovens and Murray Advertiser*, 10 August 1901: 7c; *Age* 14 June 1906: 8c; *Wodonga and Towong Sentinel* 31 October 1902: 2e.
10 *The Age*, 17 April 1907: 10e.
11 For e.g. *The Age*, 22 May 1906: 6g; *Alexandra and Yea Standard and Yarck, Gobur, Thornton and Acheron Express*, 6 May 1910: 3a; *Bairnsdale Advertiser and Tambo and Omeo Chronicle*, 4 July 1899: 2g–h; 24 May

1904: 2d; 18 April 1907: 2h; *Gippsland Times*, 23 October 1899: 3b; *Seymour Express and Goulburn Valley Avenel, Graytown, Nagambie, Tallarook and Yea Advertiser*, 13 Oct 1917: 3b.

12 *Ovens and Murray Advertiser*, 3 March 1894: 6e.

13 Birrell, R. 1998, *Staking a Claim: Gold and the Development of Victorian Mining Law*, Melbourne University Press, Melbourne, 135.

14 Hansard vol. 105, 22 October 1903, 854.

15 For e.g. *Ovens and Murray Advertiser*, 10 Oct 1903: 10g; *Mount Alexander Mail*, 24 Oct 1903: 2d.

16 Hansard vol. 105, 4 November 1903, 1063; 1065.

17 *Ovens and Murray Advertiser*, 14 November 1903: 8a–b.

18 *Ovens and Murray Advertiser*, 14 November 1903: 8a.

19 *The Argus*, 10 November 1903: 8c–d.

20 Hansard vol. 105, 8 October 1903, 515.

21 Hansard vol. 105, 20 October 1903, 773; 855; 971.

22 Hansard vol. 105, 20 October 1903, 772.

23 Hansard vol. 105, 29 October 1903, p.1018; 1019; 1042.

24 Hansard vol. 105, 28 October 1903, 969; 855; 931.

25 Hansard vol. 105, 4 November 1903, 1058.

26 Hansard vol. 105, 21 October 1903, 773.

27 Hansard vol. 105, 21 October 1903, 775.

28 Hansard vol. 105, 27 October 1903, 912.

29 Hansard vol. 105, 4 November 1903, 1050.

30 Hansard vol. 105, 4 November 1903, 1050.

31 Hansard vol. 105, 28 October 1903, 969.

32 Hansard vol. 105, 22 October 1903, 848; John Fletcher held numerous public positions at Beechworth over the years and served as a mine manager and on the district Mining Board. He was MLA for Bogong between 1902 and 1904.

33 Hansard vol. 105, 27 October 1903, 913.

34 Hansard vol. 105, 27 October 1903, 927.

35 Hansard vol. 105, 27 October 1903, 921.

36 Hansard vol. 105, 22 October 1903, 828.

37 Hansard vol. 105, 27 October 1903, 921.

38 Hansard vol. 105, 27 October 1903, 923; 1045; 1050.

39 Hansard vol. 105, 4 November 1903, 1040.

40 Hansard vol. 107, 20 July 1904, 340.

41 Hansard vol. 107, 9 August 1904, 716.

42 Hansard vol. 108, 24 August 1904, 1215.

43 Hansard vol. 108, 24 August 1904, 1211.

44 Hansard vol. 108, 24 August 1904, 1211.

45 Hansard vol. 108, 24 August 1904, 1214.

46 *Commonwealth of Australia Constitution Act* 1900, section 100.

47 Keating, J. 1992, *The Drought Walked Through*, 71–87; Gergis, J.
    2018, *Sunburnt Country: The History and Future of Climate Change
    in Australia*, Melbourne University Press, Melbourne, 91–94.
48 *The Age,* 14 June 1906.
49 Davis, J., Rutherfurd, I. and Finlayson, B. 1997, Reservoir Sedimentation
    Data in South-Eastern Australia, *Water* March/April: 11–15.
50 Hansard vol.108, 24 August 1904, 1217.
51 Greenland, P. 2001, *Hydraulic Mining in California: A Tarnished Legacy*,
    The Arthur H. Clark Company, Spokane, 263.
52 Hansard vol. 105, 27 October 1903, 916.
53 Hansard vol. 107, 20 July 1904, 341–343; *Mines Act* 1904, section 60.
54 *Ovens and Murray Advertiser*, 28 January 1905: 2.
55 *Mount Alexander Mail*, 9 June 1905: 2d.
56 *Mount Alexander Mail*, 22 July 1905: 3b.
57 *Bendigo Advertiser*, 9 September 1905: 5e.
58 *Bendigo Advertiser*, 9 September 1905: 5e.
59 *The Age*, 16 September 1905: 12f.
60 *The Age*, 11 December 1905:6e, 19 January 1906: 8b.
61 *Mount Alexander Mail*, 2 February 1906: 2d.
62 *The Age*, 14 June 1906: 8c.
63 *The Age*, 16 June 1906: 9a–b.
64 *The Age*, 3 April 1906: 4d.
65 *Report of the Sludge Abatement Board* 1906: 76.
66 *The Age*, 4 April 1906: 9c.
67 *The Age*, 4 April 1906: 9c.
68 Hansard 5 December 1907: 3009.
69 Hansard 5 December 1907: 3009.
70 Hansard 5 December 1907: 3016; 3017; 3026; 3133.
71 Hansard 5 December 1907: 3027.
72 *Report of the Sludge Abatement Board* 1906: 69.
73 *Report of the Sludge Abatement Board* 1907: 71.
74 *Report of the Sludge Abatement Board* 1906: 69.
75 *Report of the Sludge Abatement Board* 1906: 69; 1907: 71; 1908: 70.
76 *Report of the Sludge Abatement Board* 1907: 72–76; 1908: 71.
77 *Report of the Sludge Abatement Board* 1909: 68.
78 *Report of the Sludge Abatement Board* 1910: 71.
79 *Report of the Sludge Abatement Board* 1910: 71.
80 *Report of the Sludge Abatement Board* 1911: 68.
81 *Report of the Sludge Abatement Board* 1915: 61.
82 *Wangaratta Chronicle* 14 February 1914: 3g; *Ovens and Murray
    Advertiser*, 12 June 1918: 4a; *The Age*, 26 January 1934: 9e.
83 For e.g. *The Age*, 12 March 1913: 8h.
84 Dredging and Sluicing Inquiry Board 1914: 11.

85   Dredging and Sluicing Inquiry Board 1914: 14; 15; 29.

86   Dredging and Sluicing Inquiry Board 1914: 11.

87   For e.g. *The Argus*, 23 January 1914: 9c; 25 March 1914: 8i; 2 July 1914: 8h; *The Age*, 17 June 1914: 10c; *Omeo Standard and Mining Gazette*, 31 March 1914: 3a; *Ovens and Murray Advertiser*, 24 January 1914: 4d; *Wangaratta Chronicle*, 28 January 1914: 2f; 3f.

88   *Mining By-Laws of and for the State of Victoria*, 1916. Government Printer, Melbourne, 10.

89   Sludge Abatement Board 1913, *Report of the Sludge Abatement Board for the Year 1912*, Plate XXVII.

90   Lloyd, B. 2006, *Gold in the North-East: A History of Mining for Gold in the Old Beechworth Mining District of Victoria*, Histec Publications, Hampton East, Victoria, 156.

91   J. Bowser, Victoria, *Parliamentary Debates*, Legislative Assembly, 1 July 1914, 104.

7 Aftermath

1   For e.g. Lister, T.E. 2014, 'Recovery of critical and value metals from mobile electronics enabled by electrochemical processing', *Hydrometallurgy* 149: 228-237; 'Mining Your iPhone', 911 Metallurgist website, https://www.911metallurgist.com/mining-iphones/, accessed 17 August 2018; Bianca Nogrady, 'Your phone is full of untapped precious metals', BBC website, http://www.bbc.com/future/story/20161017, accessed 17 August 2018; Frank Piasecki Poulsen, 'Children of the Congo who risk their lives to supply our mobile phones', *The Guardian* website, https://www.theguardian.com/sustainable-business/blog/congo-child-labour-mobile-minerals, accessed 17 August 2018; 'History of the Super Pit' Kalgoorlie Consolidated Gold Mines website, https://www.superpit.com.au/about/history/, accessed 17 August 2018.

2   Doig, T. 2015, *The Coal Face*. Penguin, Melbourne; 2019, *Hazelwood*, Penguin, Melbourne.

3   'Exploration and mining rehabilitation', NSW Department of Industry website, http://www.nswmining.com.au/environment/rehabilitation-mine-closure, accessed 15 August 2018.

4   Campbell, R., Linqvist, J., Browne, B., Swann, T., and Grudnoff, M., 2017, *Dark Side of the Boom (NSW): What We Do and Don't Know About Mines, Closures and Rehabilitation in New South Wales*, Australia Institute, Canberra, 9.

5   'Mining Legacies', Mineral Policy Institute website, http://www.mininglegacies.org, accessed 15 August 2018; 'Savage River Rehabilitation Project', EPA Tasmania website, https://epa.tas.gov.au/epa/water/remediation-programs/savage-river-rehabilitation, accessed 15 January 2018.

6   'Legacy Mines Program', NSW Department of Planning & Environment
     website, https://www.resourcesandgeoscience.nsw.gov.au/landholders-
     and-community/minerals-and-coal/legacy-mines-program, accessed 24
     June 2019; Unger, C., Lechner, A., Glenn, V., Edraki, M. and Mulligan,
     D.R. 2012, 'Mapping and Prioritising Rehabilitation of Abandoned
     Mines in Australia', Life-of-Mine Conference, Brisbane, 262.
7   Unger et al. 2012: 262, 'Mining Legacies', Mineral Policy Institute
     website, http://www.mininglegacies.org, accessed 15 January 2018.
8   Northern Territory 2013, New South Wales 1974, Queensland
     2001, South Australia 2004, Tasmania 1995, Western Australia
     2013; 'Mining Legacies', Mineral Policy Institute website,
     http://www.mininglegacies.org, accessed 15 January 2018.
9   'Rehabilitation', NSW Department of Planning & Environment website,
     www.resourcesandenergy.nsw.gov.au/miners-&-explorers/codes-of-
     practice/rehabilitation, accessed 4 September 2018; 'Mining Legacies',
     Mineral Policy Institute website, http://www.mininglegacies.org,
     accessed 15 January 2018.
10   Unger et al. 2012: 262.
11   Campbell et al., *Dark Side of the Boom*, 1.
12   Campbell et al., *Dark Side of the Boom*, 11.
13   'Mining Legacies', Mineral Policy Institute website, http://www.
     mininglegacies.org, accessed 15 January 2018; Environmental Defenders
     Office 2012, *Reforming Mining Law in Victoria*, Environmental
     Defenders Office (Victoria), Melbourne, 20.
14   Condon, E.J. 1969, *Annual Report of the Mines Department Victoria for
     the Year Ended December 31*, 1969, Mines Department, Melbourne, 21.
15   Context 2007, *Victorian Water Supply Heritage Study: Volume 1,
     Thematic Environmental History*, Heritage Victoria, Melbourne
     p. 19; Grimshaw, P. and Fahey, C. 1982, 'Family and Community in
     Nineteenth-Century Castlemaine', *Australia 1988 Bulletin* No. 9, 88–125;
     Russell, G. 2009, *Water for Gold! The Fight to Quench Central Victoria's
     Goldfields*, Australian Scholarly Publishing, Melbourne, 137; 208.
16   Dja Dja Wurrung Clans Aboriginal Corporation 2014, *Dhelkunya Dja:
     Dja Dja Wurrung Country Plan 2014–2034*, Dja Dja Wurrung Clans
     Aboriginal Corporation, Bendigo.
17   'Human Rights to Water and Sanitation', 'Water Quality and
     Wastewater', United Nations Water website, http://www.unwater.org/
     water-facts/human-rights/, accessed 26 November 2018.
18   Satur, P. and Batagol, B. 2018, 'City Water Restrictions Hurt Our Most
     Vulnerable – Especially Women', *The Conversation*, 22 November 2018.
19   Powell, J. 1989, *Watering the Garden State: Water, Land and Community in
     Victoria 1834–1988*, Allen and Unwin, Sydney, 146; Yule, V. 1974, 'Murray,
     Stuart (1837–1919)' *Australian Dictionary of Biography* vol. 5, 322–323.

20 Mulligan, M. and Hill, S. 2001, *Ecological Pioneers: A Social History of Australian Ecological Thought and Action*, Cambridge University Press, Cambridge, 12–13; 195–196; Yeomans, P.A. 1958, *The Challenge of Landscape: The Development and Practice of Keyline*, Sydney, Keyline Publishing Ltd.

21 Mulligan and Hill, *Ecological Pioneers*, 195.

22 Mollison, B. 1988, *Permaculture: A Designers' Manual*, Tagari Publications, Sisters Creek, Tasmania, 56; 156; 413.

23 Clark, S.D. 1971, 'Australian Water Law – An Historical and Analytical Background', PhD thesis, University of Melbourne, 62–64.

24 Clark, 'Australian Water Law', 162–163; Cabena, P.B. 1983, *Victoria's Water Frontage Reserves: An Historical Review and Resource Appreciation*, Department of Crown Lands and Survey.

25 Nathan, E. 2007, *Lost Waters: A History of a Troubled Catchment*, Melbourne University Press, Melbourne; Wright, R. 1989, *The Bureaucrats' Domain: Space and the Public Interest in Victoria, 1836–1884*, Oxford University Press, Melbourne.

26 Cabena, *Victoria's Water Frontage Reserves*, 34; 69.

27 Wright, *The Bureaucrats' Domain*, 224.

28 Cabena, *Victoria's Water Frontage Reserves*, 8; Department of Environment, Land, Water and Planning 2016, *Catchment Management Authorities Fact Sheet 2*, Melbourne.

29 Nathan, *Lost Waters*, 74–78; Biologist Garrett Hardin famously identified 'The Tragedy of the Commons' in a 1968 paper in *Science* 162(3859): 1243–1248.

30 Stoeckel, K., Webb, R., Woodward, L. and Hankinson, A. 2012, *Australian Water Law*, Thomson Reuters, Sydney, 11.

31 Wahlquist, A., 2008, *Thirsty Country: Options for Australia*, Jacana Books/Allen and Unwin, Sydney, 143–179.

32 'Aboriginal People and Water', National Cultural Flows Research Project website, http://www.culturalflows.com.au, accessed 18 February 2019.

33 Australian Bureau of Agricultural and Resource Economics and Sciences, *Australian Water Markets Report 2016–2017*, Canberra.

34 For e.g. Godden, L. 2008, 'Property in Urban Water: Private Rights and Public Governance', in P. Troy (ed.), *Troubled Waters: Confronting the Water Crisis in Australia's Cities*, ANU E Press, Canberra, 157–185.

35 Powell, *Watering the Garden State*; Stoeckel et al., *Australian Water Law*, 65.

36 Clark, 'Australian Water Law', 141.

37 La Nauze, J.A. 1965, *Alfred Deakin: A Biography*, Melbourne University Press, Melbourne, vol. 1, 7.

38 Clark, *Australian Water Law*, 141.

39 For e.g. Harris, E. 2006, 'Development and Damage: Water and Landscape Evolution in Victoria, Australia', *Landscape Research* 31(2): 169–181;

Tisdell, J. 2014, 'The Evolution of Water Legislation in Australia', in K.W. Easter and Q. Huang (eds), *Water Markets for the 21st Century: What Have We Learned?*, Global Issues in Water Policy 11, Springer, Dordrecht, 164–165.

40 Deakin, A. 1885, *Royal Commission on Water Supply. First Progress Report. Irrigation in Western America*, No.19, Parliament of Victoria, Melbourne, 55. This interpretation is supported by Sandiford Clark, an authority on Australian water law, who writes that 'It is here that we first see the State acting in the capacity of a grantor of water rights, and this idea eventually pervaded most Victorian legislation relating to water.' See Clark *Australian Water Law*, 153; 184.

41 For e.g. *The Weekly Times*, 28 November 2018: 9.

42 Marshall, V. 2017, *Overturning Aqua Nullius: Securing Aboriginal Water Rights*, Aboriginal Studies Press, Canberra; Weir, J. 2009 *Murray River Country: An Ecological Dialogue with Traditional Owners*, Aboriginal Studies Press, Canberra.

43 For e.g. *The Age*, 18 July 2015; Supreme Court of Victoria, Common Law Division, Stanley Rural Community v Stanley Pastoral Pty Ltd, 12 December 2016.

44 Easter, K.W. and Qiuqiong, H. 2014, 'Water Markets: How Do We Expand Their Use?' in K.W. Easter and Q. Huang (eds), *Water Markets for the 21st Century: What Have We Learned?*, 1–9.

45 'Mineral licences', Victoria State Government Earth Resources website, http://earthresources.vic.gov.au/earth-resources-regulation/licensing-and-approvals/minerals, accessed 12 September 2018.

46 'Rehabilitation', Victoria State Government Earth Resources website, https://earthresources.vic.gov.au/community-and-land-use/environment/rehabilitation, accessed 12 September 2018.

47 Earth Resources Regulation Victoria 2017, Technical Guideline: Design and Management of Tailings Storage Facilities.

48 Environmental Defenders Office (Victoria) 2012, *Reforming Mining Law in Victoria*, 16–17.

49 For e.g. Government of Australia 2016, *Tailings Management: Leading Practice Sustainable Development Program for the Mining Industry*, https://www.industry.gov.au/data-and-publications/leading-practice-handbooks-for-sustainable-mining, accessed 12 September 2018; *From Start to Finish: A Life of Mine Perspective*, AusIMM; Davis, C. 2015, 'Environmental Rehabilitation and Mine Closure', *AusIMM Bulletin* February 2015, https://www.ausimmbulletin.com/feature/environmental-rehabilitation-and-mine-closure/.

50 'Leading Practice Handbooks for sustainable mining', Australian Government Department of Industry, Innovation and Science website, https://archive.industry.gov.au/resource/Programs/LPSD/Pages/LPSDhandbooks.aspx, accessed 12 September 2018.

51   Hamilton, J. 2015, *Adjudication on the Goldfields in New South Wales and Victoria in the 19th Century*, Federation Press, Sydney, 188.

52   Thompson, G.T. 1981, *A Brief History of Soil Conservation in Victoria – 1834-1961*, Soil Conservation Authority, Melbourne, 9.

53   Unglik, A. 1996, *Between a Rock and a Hard Place: The Story of the Development of the EPA*, Environment Protection Authority, Melbourne.

54   *Victoria Government Gazette* 60, 23 June 1971, 2169.

55   Kellow, A. and Niemeyer, S. 1999, 'The Development of Environmental Administration in Queensland and Western Australia: Why are They Different?' *Australian Journal of Political Science* 34(2): 209; Wegner, J. 2009, 'Sludge on Tap: Queensland's First Water Pollution Legislation', *Environment and History* 15(2): 206.

56   Robin Tennant-Wood, 'Sending Mines to Rehab', *The Conversation*, 1 August 2011.

57   Environmental Defenders Office (Victoria) 2012, *Reforming Mining Law in Victoria*, 21.

58   Mudd, G.M. 2013, 'Australia's Mining Legacies', *Arena* 124: 23.

59   'Mining Legacies', Mineral Policy Institute website, http://www.mininglegacies.org, accessed 12 September 2018.

60   *Sydney Morning Herald* 12 March 2018; *Bendigo Advertiser* 16 April 2018; Newcrest Mining Limited, Press Release, 10 March 2018.

61   *Ovens and Murray Advertiser*, 30 June 1877: 4–5.

62   Vines, M. 1979, 'Bowser, Sir John (1856–1936)', in B. Nairn and G. Serle (eds), *Australian Dictionary of Biography* vol. 7, Melbourne University Press, Melbourne, 365–366.

63   E.H. Cameron Victoria, *Parliamentary Debates*, Legislative Assembly, 8 October 1903, 515.

64   Dredging and Sluicing Inquiry Board 1914, *Report Upon Complaints of Injury by Dredging and Sluicing*, 12.

65   Powell, *Watering the Garden State*, 150–167.

66   Griffiths, T. 1987, *Beechworth: An Australian Country Town and its Past*, Melbourne, Greenhouse Publications, 73–94.

67   Billson quoted by John Bowser, Victoria, *Parliamentary Debates*, Legislative Assembly, 1 July 1914, 101.

68   J. Bowser, Victoria, *Parliamentary Debates*, Legislative Assembly, 1 July 1914, 103.

69   Dredging and Sluicing Inquiry Board 1914: 15.

70   Mulligan and Hill, *Ecological Pioneers*, 13–39; 136–142.

71   *Argus* 20 January 1934: 17e.

72   Robin L. 1998, *Defending the Little Desert: The Rise of Ecological Consciousness in Australia*, Melbourne University Press, Melbourne, 2–5; 25.

73   'Rivers of Gold: Understanding how gold mining shaped Victoria's history', https://rivers-of-gold.com.

74   Davies, P., Lawrence, S., Turnbull, J., Rutherfurd, I., Grove, J., Silvester,
      E., Baldwin, D. and Macklin, M. 2018, 'Reconstruction of Historic
      Riverine Sediment Production on the Goldfields of Victoria', Australia.
      Anthropocene 21: 1–15. Environmental engineer Gavin Mudd estimates
      that the total volume of tailings and waste rock from all gold mining
      across Australia is around 7200 million tonnes. As ore grades continue
      to decline, ever-greater volumes of waste will be produced to recover
      the same amount of gold. See Mudd, 'Australia's Mining Legacies'.
75   Sludge Report 1887: 39.
76   Sludge Report 1887: 63.
77   Sludge Report 1887: 37.
78   For e.g. James, L.A. 1989, 'Sustained Storage and Transport of Hydraulic
      Gold Mining Sediment in the Bear River', California, *Annals of the
      Association of American Geographers* 79(4): 570–92; Macklin, M.G. and
      Lewin, J. 2018, 'River Stresses in Anthropogenic Times: Large-scale
      Global Patterns and Extended Environmental Timelines', *Progress in
      Physical Geography* 42: 1–21; Wohl, E. 2004, *Disconnected Rivers: Linking
      Rivers to Landscapes*, Yale University Press, New Haven, 40–93.
79   Beard, D. 1979, 'Bucket Dredging in the Upper Ovens Valley', Honours
      thesis in Geography, University of Melbourne, Melbourne, 34.
80   Cargill, D. 2005, 'The Geomorphic Recovery of the Upper Ovens River
      Following the Disturbance by historical Gold Dredging'. Honours
      thesis, School of Anthropology, Geography and Environmental Studies,
      University of Melbourne, Melbourne.
81   Baldwin, D. and Howitt, J.A. 2007, 'Baseline Assessment of Metals and
      Hydrocarbons in the Sediments of Lake Mulwala, Australia', *Lakes &
      Reservoirs: Research and Management* 12: 167–174; Davis, J., Rutherfurd,
      I. and Finlayson, B. 1997, 'Reservoir Sedimentation Data in South
      Eastern Australia', *Water* March/April: 11–15.
82   Bartley, R. and Rutherfurd, I.D. 2004, 'Re-Evaluation of the Wave Model
      as a Tool for Quantifying the Geomorphic Recovery Potential of Streams
      Disturbed by Sediment Slugs', *Geomorphology* 29(6): 737–753; Pickup,
      G., Higgins, R.J. and Grant, I. 1983, 'Modelling Sediment Transport as a
      Moving Wave – The Transfer and Deposition of Mining Waste', *Journal
      of Hydrology* 60(1): 281–301.
83   Wohl, *Disconnected Rivers*, 1.
84   There are numerous contemporary accounts but for a sample see:
      Howitt, W. 1977, *Land, Labour and Gold*, Sydney University Press
      (Longman, Brown, Green, and Longmans), reprint of 1855 edition;
      Griffiths, T. (ed), 1988, *The Life and Adventures of Edward Snell: The
      Illustrated Diary of an Artist, Engineer and Adventurer in the Australian
      Colonies 1849–1859*, Angus and Robertson, Sydney; Korzelinski, S., 1979,
      Memoirs of Gold Digging in Australia, University of Queensland Press,

Brisbane; Wynn, G. 1979, 'Life on the Goldfields of Victoria: Fifteen Letters', *Journal of the Royal Australian Historical Society* 64(4): 258–268; Westgarth, W. 1857, *Victoria and the Australian Gold Mines in 1857*, Smith, Elder and Co, London.

85 For e.g. 'The Castlemaine Association of Pioneers and Old Residents' (eds) 1972, *Records of the Castlemaine Pioneers*, Rigby Limited, Melbourne; Ferguson, C.D. 1888, *The Experiences of a Forty-Niner During Thirty-Four Years' Residence in California and Australia*, The Williams Publishing Company, Cleveland; Smith, J.G. 2002, *Reminiscences of the Ballarat Goldfield*, Pick Point Publishing, Melbourne.

86 Withers, W.B. 1887, *The History of Ballarat, from the First Pastoral Settlement to the Present Time*, F.W. Niven and Co, Ballarat (facsimile of the second edition, first published 1870), p.viii.

87 Withers, *History of Ballarat*, x.

88 Annear, R. 1999, *Nothing But Gold: The Diggers of 1852*, Text, Melbourne; Hocking, G., 2000, *Gold Fever: A Rollicking History of the Australian Gold Rush*, The Five Mile Press, Melbourne.

89 Bate, W., 1988 *Victorian Gold Rushes*, McPhee Gribble/Penguin, Melbourne; Bate, W. 2001, 'Gold: Social Energiser and Definer', *Victorian Historical Journal* 72(1–2): 7–27; For e.g. Broome, R. 1984, *The Victorians: Arriving*, Fairfax, Syme and Weldon Associates, Sydney, 67–93. The same series includes a much fuller and more useful discussion of the growth of mining and its environmental problems in Dingle, A.E. 1984, *The Victorians: Settling*, Fairfax, Syme, and Weldon Associates, Sydney, 39–57, 89–101; Powell, J.M. 1976, *Environmental Management in Australia, 1788–1914*, Oxford University Press, Melbourne; Wright, *The Bureaucrats' Domain*.

90 For e.g. Fahey, C. and Mayne, A. (eds) 2010, *Gold Tailings: Forgotten Histories of Family and Community on the Central Victorian Goldfields*, Australian Scholarly Publishing, Melbourne; Reeves, K. and Nichols, D. (eds) 2007, *Deeper Leads: New Approaches to Victorian Goldfields History*, Ballarat Heritage Services, Ballarat; Roberts, G. *Metal Mining in Tasmania 1804–1914*.

91 For e.g. Knox, M. 2013, *Boom: The Underground History of Australia, from Gold Rush to GFC*, Viking, Melbourne.

92 Woodland, J., 2001, *Sixteen Tons of Clunes Gold*, John Woodland, Melbourne; Birrell, R. 2005, 'The Development of Mining Technology in Australia 1801–1945', PhD thesis, University of Melbourne; Davey, C. 1996, 'The Origins of Victorian Mining Technology, 1851–1900', *The Artefact* 1952–1962; Birrell, R., 1998, *Staking a Claim: Gold and the Development of Victorian Mining Law*, Melbourne University Press, Melbourne; Lloyd, L. 1966, 'The Sources and Development of Australian Mining Law', PhD thesis, Law, Australian National University,

Canberra; Woodland, J. 2014, *Money Pits: British Mining Companies in the Californian and Australian Gold Rushes of the 1850s*, Ashgate, Farnham, Surrey; Lloyd, B., 2006, *Gold in the North-East: A History of Mining for Gold in the Old Beechworth Mining District of Victoria*, Histec Publications, Melbourne; Lloyd, B. and Coombes, H. 2010, *Gold in the Walhalla Region, West Gippsland, Victoria*, Histec Publications, Melbourne; Woods, C. 1985, *Beechworth: A Titan's Field*, Hargreen Publishing, Melbourne.

93 Bate, W. 1978, *Lucky City: The First Generation at Ballarat*, Melbourne University Press, Melbourne; Cusack, F. 1973, *Bendigo: A History*, Heinemann, Melbourne.

94 Dunstan, D. 1996, *Victorian Icon: The Royal Exhibition Building, Melbourne*, The Exhibition Trustees in Association with Australian Scholarly Publishing, Melbourne.

95 For e.g. Lennon, J. 2000, 'Victorian Goldfields: Tentative World Heritage Listing', *Historic Environment* 14(5): 70–74; Smith, A. and Lawrence, S. 2018, 'Understanding the Outstanding Universal Value of Mining Sites: Evolving International Approaches and Their Implications for Reconsidering the World Heritage Potential of the Victorian Goldfields', *Historic Environment* 30(1): 50–63.

96 Reeves, A., McCalman, I., and Cook, A. (eds) 2001, *Gold: Forgotten Histories and Lost Objects of Australia*, Cambridge University Press, Cambridge; Wright, C. 2013, *The Forgotten Rebels of Eureka*, Text, Melbourne; Dja Dja Wurrung Clans Aboriginal Corporation 2014, *Dhelkunya: Dja Dja Wurrung Country Plan 2014–2034*, Dja Dja Wurrung Clans Aboriginal Corporation, Bendigo; Cahir, F., 2012, *Black Gold: Aboriginal People on the Goldfields of Victoria, 1850–1870*, ANU Epress, Canberra; Attwood, B. 2017, *The Good Country: The Dja Dja Wurrung, the Settlers and the Protectors*, Monash University Publishing, Melbourne.

97 Frost, W. 2011, 'Visitor Interpretation of the Environmental Impacts of Gold Rushes at the Castlemaine Diggings National Heritage Park, Australia', in Conlin, M. and Jolliffe, L., (eds), *Mining Heritage and Tourism: A Global Synthesis*, Routledge, London, 97–107; Frost, W. 2013, 'The Environmental Impacts of the Victorian Rushes: Miners' Accounts During the First Five Years', *Australian Economic History Review* 53(1): 72–90; Garden, D. 2001, 'Catalyst or Cataclysm? Gold Mining and the Environment', *Victorian Historical Journal* 72 (1&2): 28–44; McGowan, B. 2001, 'Mullock Heaps and Tailing Mounds: Environmental Effects of Alluvial Goldmining', in I. McCalman, A. Cook and A. Reeves (eds), *Gold: Forgotten Histories and Lost Objects of Australia*, Cambridge University Press, Cambridge, 85–100; Reeves, K. and McConville, C. 2011, 'Cultural Landscape and Goldfield Heritage: Towards a Land Management Framework for the Historic South-West Pacific Gold

Mining Landscapes', *Landscape Research* 36(2): 191–207; Cuthbertson, D. 2016, 'La Mama Takes over Historic Steiglitz Homestead to Stage Uncle Vanya', *The Age,* 8 April 2016. An international perspective on these issues is provided in Mountford, B. and Tuffnel, S. 2018, *A Global History of Gold Rushes*, University of California Press, Berkeley.

Conclusion

1   Trigger, D. 1997, 'Mining, Landscape and the Culture of Development Ideology in Australia', *Ecumene* 4(2): 161–180.
2   Parbhakar-Fox, A. 2016, 'Treasure from Trash', *The Conversation*, 29 June 2016; Daly, J. 'Mining's War on Waste', *ABC News* 11 April 2018.
3   'Mining Your iPhone', 911 Metallurgist website, https:// www.911metallurgist.com/mining-iphones/, accessed 17 August 2018; Bianca Nogrady, 'Your phone is full of untapped precious metals' BBC website, http://www.bbc.com/future/story/20161017, accessed 17 August 2018.

INDEX

ABC 251

Aboriginal people 71, 108–9, 216, 223
see also Dja Dja Wurrung people

Act for amending the Laws relative to the Gold Fields 1857 118

Act for the Better Management of the Gold Fields 1855 117

Act to amend the Law relating to Leases of Auriferous Lands and for other purposes 1862 126

Adamson, James 111

Adelaide, SA 38

Africa, water in 217

The Age 63, 199, 204

agriculture, sludge and
    complaints from farmers 22, 28, 138, 174
    effect on animals 25–7, 137
    effect on crops 12, 17–19, 22
    use of sludge for farming 13

Ah Chong 154

Ah Fee 99

Ah Fong 10

Ah Moon 104

Ah Nam 105

Ah Ping 104, 171

Ah Pow 104

Ah Soon 99

Ah Tan 98

Ah Tang 105

Albert Street, Creswick 144–5

Albury, NSW 181

Alma and Red Hill Mining Company 105

American Company 76

Amherst, Vic 27, 79, 85

Amos, Mark 86

Amos, Thomas 94

An Act to provide for the Disposal of Sludge from Alluvial Mines in Creswick 1885 167

Andrews, Charles 185

Annual Report on Dredge Mining and Hydraulic Sluicing 13

aqueducts *see* water races

Ararat, Vic 81, 93, 140

The Argus 70, 183–4

Argus Company 55–6

arsenic contamination 21

Ashworth, Thomas 189

Asia, water in 217

Atkins, John 157

'Australasian' mine 158, 177

Australia brig 88

Australia Institute 213

Australian Broadcasting Corporation 251

Australian Dictionary of Biography 218, 233

Australian Natives' Association 236

Avoca River 72

Baarmutha *see* Three Mile Creek
Bailes, Alfred 184–6
Bairnsdale, Vic 24
Bald Hill, Vic 95
Bald Hills, Vic 107, 154
Ballarat, Vic *see also* Eureka rebellion
  capping and filling mine shafts
    213–14
  dams 79, 202
  dredging at 186
  historical accounts 243–4
  mines 20, 25, 57, 98
  mining companies 82
  sludge in 19–20, 25, 64, 154–7,
    169, 217, 238
  sludge pottery 13
  sludge regulations 140, 143–4,
    201, 203–4
  sluicing at 76
  surveyors 128
  water regulations 131
  water supply 70–2, 81, 101, 217
Ballarat Sludge Commission 19
The Ballarat Star 154
Ballarat Water Commission 101
Balmain, Thomas 146
Baltimore ship 79
bank rights 122
Bannear, David 214
Barkers Creek 103, 146
Barwon River 21, 25
Bass Strait 7, 34, 72
Bate, Weston 243
Batesford, Vic 21
Batman, John 108
Beard, David 240
Beechworth, Vic *see also* Ovens
  goldfield
    dams 124, 170
    decline of mines 177
    drainage 97, 140, 164
    landscape 13
    sludge 25, 27–9, 122, 137, 164, 188

  sluicing at 40–1, 69, 97
  spring water 123–4
  water bosses 95
  water regulations 80–1, 117–20,
    126–8, 131–2
  water supply 66, 72, 84, 103
Beechworth Local Court 113–14, 117
Beechworth Police Court 106
Bendigo, Vic
  Aboriginal people in 246
  capping and filling mine shafts
    213–14
  Catherine Reef United mining
    company 57
  Coliban System of Waterworks
    54–5, 70
  Cornish population 42
  dams 63–4, 202
  Fortuna Villa 245
  historical accounts 244
  panning and cradling 46
  Perseverance Reef 83
  puddling and sluicing 63
  sludge at 14–19, 26, 148, 169, 173,
    201, 238
  sludge channels 140, 150–2
  water regulations 112, 131
  water supply 71–2, 82–3, 217
Bendigo Advertiser 197
Bendigo and Fryers Company 55
Bendigo Creek
  dams 83
  sludge at 12, 14–18, 148
  sludge channels 150–2, 169
Bendigo Sludge Commission 13
Bet Bet Creek 22, 65
Beveridge, John 239
Beveridge, Tom 25
BHP 2–3
'Big Ditch' 100
Billson, Alfred 31, 191, 200, 205,
    234
Birch Creek 23, 50, 165

Bird, Charles Grey 69, 86
bitumen 92–4
Black Hill, Vic 20
Black Lead, Vic 99, 145, 147
Blackman's Lead, Vic 10
Blackstone, William 115
Bladier, Jacques 12, 17, 30–1, 150
Blainey, Geoffrey 243
Bland, Rivett Henry 50
Board Appointed to Inquire into
    the Sludge Question 167–9, 171
        in Ballarat 20
        in Beechworth 28–9, 170
        in Carisbrook 22–3
        in Invermay 26
        in Maryborough 23
        in Tarrawingee 138–9
        in Wangaratta 27, 29
        in Yackanadah 30
Board of Inquiry into Complaints
    of Dredging and Sluicing 203–6,
    233–4
Bonshaw Freehold Company 157
Boorhaman, Vic 28
Bourke, Richard 108
Bowser, John
        action on sludge 179, 181–2,
            187–8, 191, 193–5, 205
        on activism 208
        on Alfred Billson 235
        background 178–9
box sluices 40 see also sluices
Boyd, Henry 105
Braché, Jacob 80
Bradfield, Arthur 92
Bradfield, Ray 92
Bragg, John Boadle 79, 92, 107
Bragg's Dam 79
Braidwood, NSW 42
Brazil 1–2, 231
Bren, Leon 74–5
Bright, Vic 25, 34
Briseis Company 230

British Empire see colonisation;
    Crown land regulations
Broken Hill, NSW 209
Brown, James Drysdale 205
Brown, Joseph 188
Browne, Gilbert 64, 150
Brumadinho, Brazil 2
bucket dredges see dredges
Buckland River 42, 91, 97, 112
Buckland Valley 62, 112
Bucknall, Albert 240
Bull, John E.N. 142
Bullarook Creek (Birch Creek) 23,
    50, 165
Bullarook Forest 70, 87, 89, 111
Bungaree, Vic 99
Burrowes, Robert 158–9, 161
Burrumbeet Creek 26
Busch, Philip 171

Cabbage Tree Flat 89
California, US
        miners from 42, 68, 76
        mining techniques 40–1, 76, 172
        sand slugs 66
        water rights 132–3
California Debris Commission 193–4
California Debris Commission Act
    1893 193
'the California Doctrine' 133
Cameron, Ewen 182, 189–90, 233
Cameron, James 191
Caminetti Act 1893 193
Campaspe River 22
Campbell, James 31
Campbells Creek 25, 103, 146, 172
Canada 87, 89
'Canada race' 87–9
Canterbury Park (Eaglehawk Public
    Gardens) 13, 169
Cargill, Deborah 241
Carisbrook, Vic 22–3, 173–4
Carpenter, Thomas 149

Castlemaine, Vic
    Aboriginal people in 246
    Chinese miners in 42, 82
    puddlers 82, 142
    sand slugs 25
    sludge channels 152
    sludge in 10, 103, 146
    sludge regulations 140, 142
    sluicing 80
    timber 57
    water supply and regulations
        53–5, 58, 112
Castlemaine Diggings National
    Heritage Park 245
Castlemaine Dredging Association
    197–8
Castlemaine Mining Board 233
Catchment Management Authorities
    247
Cathcart, Michael 114
Catherine Reef United mining
    company 57
Cesar Godeffroy ship 38
Chamber of Mines 198
channelising see also water races
    of Bendigo Creek 150–2, 169
    of Creswick Creek 144–5, 152
    of Fryers Creek 146, 152
    of Hodgson Creek 164–5
    legislation 158–9, 167
    of Loddon River 61
    in Tarrawingee 177
    of Yarowee Creek 144
Chewton, Vic
    sludge in 2–3, 103, 146, 173
    sluicing 173
    waterwheels 56
child mortality 217
Chin, Wai Jung 99
China, minerals from 209
Chinese residents
    conflicts 104–6, 112–13
    deaths of 46

panning and cradling 46
puddling 82
sludge and 145–7
sluicing 42, 46
use and control of water 98–9, 216
Clarke, Alexander 25, 196–7, 201, 203
Clunes, Vic
    decline of mines 177
    mining companies in 49–50,
        125, 158, 165, 169
    sludge in 22–3, 174
    water supply 165
coal mines 210
Cocks Eldorado Dredge 59–61
Cocks Pioneer Gold and Tin Sluicing
    Company 27, 59–61
Coliban System of Waterworks
    cost of 57, 162
    issues with 70
    sludge and 103, 171–2
    structure of 54–5, 94–5
colonisation 71, 108–10
Columbia Ditch 41
Confidence Dredge 205
Connolly & Co 124
conservation 235–6
Constitution, Australian 192
Cooper, George 55
Cope, Thomas 118–19, 126
Cornwall, UK 40–2, 118
Corowa Progress Association 181
Córrego do Feijão iron mine 2
Council of Australian Governments
    224
Court of Mines 229
cradling and panning 46
Craig's Hotel 166
creek rights 122
Creswick, Vic
    conflicts 105–7, 110–12
    dams 77–9
    decline of mines 177
    floods 152

sludge flushing 203
sludge in 22–3, 25, 31–2, 50, 158, 161
sluicing 43, 100
water supply 23, 71, 87–9, 99, 102, 110–12
Creswick Advertiser 89
Creswick Court of Mines 106
Creswick Creek
  landscape 13
  sludge in 145–6, 169
  sluicing at 99
  structure of 50, 76–7, 79, 144–6, 152
Creswick State Forest 5, 74, 78
Crown land regulations 108–10, 118–19, 124–6, 221–2

dams *see also* Eatons Dam
  in Ballarat 202
  in Bendigo 72, 82–3
  Bragg's Dam 78
  in Brazil 1–2, 231
  dredging in 62–4
  flooding 1–2, 231
  keyline water management 219–20
  Lake Kerferd 124
  at Mount Morgan 231
  at Newstead 62
  practice of using 62–4
  regulations 116, 122, 131, 138, 159, 167, 174, 207, 228, 230
  sluicing and 68, 80–1, 116, 170, 207
  structure of 74–5, 78–80
  on Three Mile Creek 138
  on Yackanadah Creek 68
  Yankee Dam 76–7
  on the Yarra River 95
Dargo River 23
Darling River 71, 222–4
Davidson, Henry 29, 128

Daylesford, Vic
  sludge in 14, 17, 25
  water races in 85–6, 100
de Lima, Jacinto 79, 105–7
Deakin, Alfred 176, 224–5
Deason, John 42
Deep Creek *see* Tullaroop Creek
Department of Mines 46, 49
depressions, economic 177, 219
Derelict Mines Program 212
Devil's Gully Tunnel 94–5
diamonds 49
diarrhoea 217
Dicker's Mining Record 70, 77
Dja Dja Wurrung people 152, 217, 246
Dolly's Creek 101, 102
Dow, William 10
Dowling Forest 26
Drainage of Mines Act 1887 48
dredges
  Cocks dredge 59–61
  damage caused by 32–4, 61–3, 180–1, 188, 196–200
  Newstead dredge 62
  numbers of 179, 206
  process of using 60, 180
  regulations 62–3, 190, 205, 207–8
  water use 60, 62
Dredging and Sluicing Inquiry Board 13, 33, 235
drinking water 23–4, 216–17
droughts 67, 82, 187, 217 *see also* Federation Drought
Dublin, Ireland 41
Duffy, John Gavan 163
Duncan, William 9
Dunnolly goldfields 109
dysentery 217

Eaglehawk, Vic 26
Eaglehawk Public Gardens 13, 169

Eaton, Benjamin Franklin 76, 110, 132
Eaton, Charles Lafayette 76–7
Eaton, Maud 77
Eaton family 110
Eatons Dam 72–8
Ebbott, John 57–8, 171
economy 3, 177, 234–5
Eddington, Vic 103, 172
Eggers, Gustave 104
Eggers, Julius 104–6
Eldorado, Vic 27, 33, 59–61
electronics 209–10, 251–2
Elmore, Vic 18
Elphinstone, Vic 55
English law 114–15
Environment Protection Authority 229
Environmental Defenders Office 230–1
Environmental Effects Statements 230
Epsom, Vic 12, 15, 30
Epsom Hotel 18
Epsom Sludge Committee 146
Eureka rebellion 117–18, 142, 246
Eureka Reef 67
Euroa, Vic 188
Eurobin, Vic 205
Evansford, Vic 79
Evelyn Tunnel Gold Mining Company 95
e-waste 251–2
Excelsior Company 170

Fairbarn, John 132
Farnsworth, Asa G. 79, 85
Federation Drought 52, 192
Federation University 21
First World War 205
Fisken, Archibald 101–2
Fitzgerald, Nicholas 164
Fletcher, Donald 119, 121, 170, 191

Fletcher, John 28, 184, 188
flooding
    in Castlemaine 10
    of Creswick Creek 145
    of dams 24–5, 73–6, 78, 231
    of Forest Creek 146
    in Inverleigh 154–5
    of mines 47
    of Tullaroop Creek 240
flumes 90–1, 94, 102, 121
Foletti, Pietro 98
Forbes, Thomas 82–3
Ford, R.G. 168
Forest Creek 2–3, 15, 103, 146
forestry 57–8
Fortuna Villa 245
Fosterville Gold Mine 231–2
Francisco, Domingo 79, 107
Fryers Creek 10, 94, 103, 172
Fryers Creek Sluicing Company 90–5
Fryerstown, Vic 43, 94, 171

Garfield waterwheel 53–8
Gaylard, Samuel 124
Geelong, Vic 11, 14, 21, 64
Geelong Advertiser 11
Geelong Anglers' Club 21
gemstones 49 see also quartz mines
gender 216–17, 246
Geoscience Victoria 214
German Gully, Vic 55
German miners
    conflicts 104–6, 112–13
    on the Ovens goldfield 38
    in Three Mile Valley 104–6, 112–13
Germano mine 1
Gillies, Duncan 176
Gillies, James 170
Gippsland, Vic 23, 48, 51, 72, 132
Glennon, John 85
Gold Fields Amending Act 1862 131–2
Gold Fields Amendment Act 1875 162

Gold Fields Commission of Enquiry 117, 142
Gold Museum 244
Golden Gully, Vic 46
Goldsmith, Len 100
Goodman, David 245
The Goulburn Herald 41
Goulburn River 71–2, 192
Goulburn Valley 24
Goulburn Weir 192
Graham, Thomas 33
Granite Flat, Vic 100
Graves, James 188, 189
Great Dividing Range 71–2
Great Rand Company 52, 58, 93
Griffiths, Tom 5
Grose, Walter 189
ground sluices 39–40 *see also* sluices
groundwater 35–6, 123–4, 226
G.S.G. Amalgamated 207
Guildford, Vic 32

Hamilton, Charles 186–7
Hamilton, William 10
Hanney waterwheel 52
Happy Valley Creek 97
Harrietville, Vic 61
Harris, Albert 183
Hazelwood, Vic 210
He King 105
Heathcote, Vic 9, 25, 86
Hedley, Thomas 9–10
Heritage Rivers Act 1992 215
Heritage Victoria 214
Hickford, Frederick 183
Higgins, Joseph 123
Hitchcock, A.F. 112
Hobsons Bay, Vic 70
Hodges, Charles P. 82
Hodgson Creek
    channelising 164–5
    sludge in 28–9, 44, 103, 136–9, 169
Holmgren, David 220

House, James C. 239
Howitt, William 242
Howqua Hills 52, 58
Howqua River 51–2, 94
Howqua United Mining Company 52
Humbug Hill Sluicing Company
    employees 79, 106–7
    practices 43, 79–80, 92–3
Humbug Hill, Vic 87, 99
Hunt, George 132
Huntly, Vic 18, 30, 150, 169
Hurdle Flat, Vic 118
hydraulic sluicing
    dangers of 44–6
    John Pund's 43–4, 84
    John Wallace's 96
    process of 44, 58–9
    regulations 190, 207
    requirements 80
    Rocky Mountain Extended Gold
        Sluicing Company's 96
    sludge produced 172

Independence Quartz Mining
    Company 82
Indigenous Australians *see*
    Aboriginal people
Inglewood, Vic 71
Inverleigh, Vic 19, 154–7
Invermay, Vic 26
iPhones 209
irrigation 192–3, 218–20
Irrigation Act 1886 225

Jim Crow Creek 13
Joyces Creek 196
J.W. Robertson & Co 89

Kalgoorlie, WA 209
Kangaroo Creek 85–6
Kangaroo Flat, Vic 150
Kassebaum, Friedrich 170
Keating, John 107

Kelly, William 15
Kerang, Vic 23
Kerferd, George Briscoe 124, 159, 163, 166
Kerstein, Henry 104
keyline water management 219–20
Kiandra, NSW 42
Kiewa River 103
Kiewa Valley 177
Kiewa Valley Anti-Dredging League 63, 181
King, He 105
Kirk, John 76
Kirks Reservoir 79
Kirkwood, Hay 186
Kirton, Joseph 186
Koch, Henry 83
Koh-i-Noor 57
Kulin nation 108

Laanecoorie Weir 24, 103, 192–3, 198
Lake, Thomas 106–7
Lake Connewarre 21
Lake Kerferd 124
Lake King 24
Lake Sambell 96
Lal Lal Creek 101
Lal Lal Waterworks Association 100–2, 132
Land Acts 109, 130
Lands Compensation Statute 1869 160
Lands Department 221
Lansell, George 245
Larritt, Richard William 149
Latrobe Valley 210
Lawrence, Susan 51–3
Lawson, Harry 198, 200
The Leafy Tree 73
legacy mines 211–15, 231
legal issues see water rights
Leigh River, sludge in 19–21, 25, 64, 155, 169

Lewis, Charles 107
Lewis, William 10
LiDAR 37–8, 91, 95
Lightning Creek flume 90
Lima, Jacinto de 79, 105–7
Lincoln Hill, Vic 87
Lindsay, Daryl 73
Little Bendigo, Vic 26
Little Desert, Vic 237
LLWA (Lal Lal Waterworks Association) 100–2, 132
Loddon Anti-Sludge Association 63, 198
Loddon Gold Dredging Company 196–7
Loddon River
    channelising 61
    flow 72
    irrigation 192–3
    overuse 215
    sludge in 23, 25, 27, 103, 169, 171–2, 180, 199, 203, 239–40
    water supply 43, 91–2, 94–5
    waterhole 239–40
Loddon Valley 24, 32, 201
Long, Tack 104–6
Long Dick's Gully 68
Long Gully 83
long toms see box sluices
Long Tunnel Extended Gold Mine 24
Lord Nelson Mine 26
Lowe Kong Meng 98
Lower Staghorn dredge 62

Madam Berry mine 31–2, 158, 177
Malaya 42
Maldon, Vic 10, 60
Malmsbury, Vic 54–5, 58
Mammoth Hydraulic Sluicing and Gold Mining Co. 90
Manchester Reef Waterwheel Company 55

Mansfield, Vic 52
Martin, Richard 46
Maryborough, Vic
    sludge in 22–3, 140, 157, 169
    water supply 81
Maryborough and Dunolly Advertiser
    157
Mason, George 110–11
Mason's Gully, Vic 99
Mayes, Charles 89
McColl, Hugh 163
McCormish, James 105
McGeorge, J.H.W 62
McIntyre, Joshua 159
McIvor Creek 10, 22, 86
McIvor Hydraulic Sluicing Company
    9, 86, 100
McKenzie, Murdock 31–2
McLeod, Donald 190, 193, 196,
    199–200, 233
Mead, Elwood 233–4
Melbourne, Vic
    bitumen production 92–3
    infrastructure 3, 245
    legislative meetings 166
    water supply 70–1, 188, 217
Melbourne International Exhibition
    245
Meng, Lowe Kong 98
mercury contamination 21
Mia Mia Creek 46
Millennium Drought 67
Miller, Catherine 98
Mine Owners' Association 166
Miners' Association 23, 166
Mines Acts
    effect of 140, 143
    formulation of 14, 182–95,
        199–201
    Queensland legislation based
        on 229
Mines Amending Act 1862 143, 153,
    206, 220

Mines Department 6, 176, 213–14, 238
Mines Inspectors 129, 153, 175
mining *see also* dredges; legacy
    mines; panning and cradling;
    puddlers; sluices
    decline of 177
    surveyors 128–9
Mining Boards 143–4, 153, 186
Mining on Private Property Act 1884
    125
mining registrars 129
Mining Statute 1865 131–2, 153, 157,
    220, 229
Mitchell, William 111–12
Mitchell River 23, 191, 201
Mitta Mitta River 30, 177, 201
Mitta Mitta, Vic 45, 90, 100, 173
Molesworth, Robert 124, 156
Mollison, Bill 220
Mollison, Crawford 63
Monash University 18
Moorabool River 14, 21, 101
Mornington Peninsula 77
Morphy, John 115–16
Morres, Henry 20, 155–6
Morrisey, John 187, 189, 191
Morrisons, Vic 21
motive power rights 122
Mount Alexander *see* Castlemaine
Mount Buffalo Chalet 234
Mount Buller, Vic 51
Mount Egerton, Vic 157
Mount Isa, QLD 209
Mount Lyell, Tas 209
Mount Morgan, QLD 231
Moysen, Alfred 118–19
Moysen v Hopper 118–19
mullock 13
Murphy, Lawrence 171
Murray, Stuart 218
Murray, Virginius 117
Murray River 7, 28, 34
Murray–Darling 71, 222–4

Myers Creek 169
Myrtleford, Vic 97

Nagambie, Vic 24, 192
National Library of Australia 6, 139
national parks 236
National Water Initiative 224
Nerrina, Vic 26
New Australasian mine 47
New Brunswick, Canada 88
New South Wales
    legacy mines 211–12
    legislation 116, 228–9
    Select Committee on Murray
        River Navigation 77
    sluicing in 49
New South Wales Sludge Abatement
    Board 229
New Zealand
    gold rush 89–90
    legislation 162
    mining techniques 32, 61,
        185–6
Newbridge, Vic 13
Newcrest mine 231
Newstead, Vic 25, 61–2, 66
Nine Mile Creek
    mining at 38, 68
    structure and water supply of
        68–9, 83–4, 103, 115
Nintingbool Sluicing Company 82
Nolan, Hopton
    career 97, 138
    sludge protests 29, 138–9, 165,
        195, 232
North Ovens Shire, Vic 29, 164–5
NSW see New South Wales
Nuggetty Gully, Vic 145

Oates, Richard 42
O'Brien, Terence 26
Ohio, US 76
O'Keefe, Edward 149

O'Loghlen, Bryan 162, 176
Omeo, Vic 45
One Mile Creek 171
Orange, NSW 209
O'Reilly, Julia 26
orphaned mines 211–15, 231
Otago, NZ 89
Outtrim, Thomas 32
Ovens and Murray Advertiser 27, 32,
    155, 183, 196
Ovens Anti-Pollution Association
    33, 63
Ovens Gold Fields Water Company
    123
Ovens goldfield
    minerals of 48–9
    Scottish emigrants on 96
    water supply 83, 120, 122
Ovens River
    channelising 164
    dredging 32, 62, 186, 188, 205
    flow 72
    heritage 215
    Hopton Nolan's land on 138
    overuse 215–16, 240–1
    sludge in 25, 28–9, 33, 169,
        178–80, 183–4, 188, 235
    water supply 97, 103, 115
Ovens River Frontage Anti-Sludge
    Pollution Association 63, 203
Ovens Valley 38, 177, 233–4
Oxley, S.J. 203
Oxley, Vic 178

panning and cradling 46
Panton, Joseph 148
Parfitt, Henry 28, 176, 195
Patent Bitumenized Pipe Company
    93
Patterson, James Brown 159
Peacock, Alexander 185, 188–9
Pendergast, James 122
permaculture 220

Perseverance Reef 83
Perth, WA 217
Peterson, Jim 18
Peterson, Lynette 18
Phillips, John 42
phones 209, 252
Pink Cliffs Geological Reserve 9
Pioneer Claim 30, 45
Plenty River 188
Pocahontas Gold Mining Company 27
Porcupine Flat, Vic 10, 60
Port Phillip and Colonial Gold Mining Company
   land ownership 125
   sludge production 177
   timber use 57
   water use 23, 49–50, 66, 165
Port Phillip, Vic 7, 108
Portuguese Flat, Vic 77, 99
Pound Bend, Vic 95
Pratt, James 28
precious stones 49
Prendergast, George 186
Prince of Wales Company 157
Public Works Office 147
puddlers
   Chinese miners' use of 82
   design of 14–15
   inefficiency 47
   legislation 142–3, 148, 150
   sludge and 63, 142–3, 147–8, 150
   water use 47
Pund, John Martin Dietrich
   career 38
   employees 96
   settling dam 207
   sluicing 43–4, 66, 170, 205, 207
   sourcing water 83–5, 103, 120, 123, 221
Pund, Percy 138, 205, 207, 230
Pund & Co 85

quartz mines
   process of mining 48, 82–3
   sludge 169, 171, 173–5, 201
Quat Quatta 245
Queensland 42, 49, 228–9
Queenstown, NZ 89–90
Quinn's Brewery 89

race 98, 216 *see also* Chinese residents
races *see* water races
Ramsar Convention 21, 65
Ramsay, Thomas 13
Red Hill, Vic 91, 95
Reed, Joseph 25
Reedy Creek 33, 60, 169
Reedy Lake 21, 65
Reform reef mine 97
Regulation of Mines Statute 1873 129
Reilly, John Henry
   sourcing water 41, 54–5, 68–70, 72, 83, 86, 115–16
   on water consumption 66
Renaissance Company 55
Richardson, Richard 163
Rio Paraopeba River 2
Rippon Lea Estate 245
River Murray Commission 224
Rivers of Gold project 238, 240
Roads Board 143, 150
Robertson, James William
   career 89
   water races 87–9, 111–12, 242
Robertson, William 156
Rochester, Vic 18
Rocky Mountain Extended Gold Sluicing Company. 96–7, 170
Rodda, James 26
Rose, William 85
Royal Commission on Gold Mining 48, 177
Royal Exhibition Building 245
Royal Mining Commission 119

Royal Sludge Commission 63, 148–50
Roycraft, John 77, 110–11, 132
The Rush that Never Ended 243
Russell, Christopher 99
Russell, George 110, 154
Russell, George, Jnr 110–11
Russell, Henrietta Ann 99
Russell family 110–11
Russells Dam 79
Rutherford, Ian 74–5

Sailor's Creek 98
Sale, Vic 24
Salter's Creek 90
Samarco's Germano mine 1
sand slugs 24–5, 65–6, 241
Sandison, James 14–17, 47, 150
Sandy Creek 30
Sankey, Richard H. 81, 85
sapphires 49
Sargood, Frederick 245
Savage River, Tas 211
Scottish miners 96
sediment *see* sludge
Select Committee on Murray River Navigation 77
settling dams *see* dams
Seymour, Vic 24
Shakespear, Robert 167–8
Sheepwash Creek 12, 46
Sheepyard Flat 51
Shelford, Vic 19, 22, 155–6, 201
Shoppee, Charles 186
siphons 90–3
Sladen, Charles 161
Slaty Creek 93, 99, 102
slimes 13
sludge
    agriculture and *see* agriculture
    deaths in 10–11, 25–7, 64
    definition of 13
    in drinking water 23–4, 216–17

production of *see* mining
regulations 136–77 *see also* Royal Sludge Commission
types of 13, 63–6 *see also* sand slugs
uses of 13–14, 122
Sludge Abatement Board
    creation of 193–6, 229
    duration 247
    legacy 229–30
    letterhead 197
    practices and impact of 20, 199–201, 203–7
Sludge Drainage Bill 158–62
sluices *see also* hydraulic sluicing
    beginning of use 41–2
    demographics of 42, 99
    Eaton family's 76
    for gemstones 49
    growth of 69
    John Pund's 38, 43–4, 66
    John Reilly's 68
    process of using 39–40, 43, 44–5, 68
    regulations 116, 122, 175
    requirements 43, 46, 66, 81
    sludge produced 63–4, 169, 173, 205
    for tin 49, 132
slums (sludge) 13
Smeaton, Vic 174
Smith, Robert Murray 160
Smith, William Collard 31, 159
Smyth, Robert Brough 46, 81, 107–8, 120, 134
Smythesdale, Vic 82
Snake Gully *see* Stanley
Snowy Mountain Hydro system 223
Soil Conservation Authority 229, 247
South Australia 212
Sovereign Hill, Vic 244
Spence, William Guthrie 23

Spencer, Edward 124
Spring Creek 17, 83, 97, 170
spring water 35–6, 123–4, 226
St Arnaud, Vic 26
St. George's Sluicing Company 93
Staffordshire Reef Quartz Mining
 Company 82
Staghorn Creek 30
Stanislaus River, US 41
Stanley, Vic
 groundwater 226
 mining at 35–6, 38, 83, 117
 sluicing at 41
 Wallace family business at 97
 water supply 36, 68, 103, 106, 120
Star Hotel and Theatre 119
State Rivers and Water Supply
 Commission 219
Sterry, David 184
Stevenson, James 43
Stewart, James Syme 79, 85
Stirling, UK 96
Sullivan, James Forester 149
Super Pit 209
surveyors 128–9
Swan Hill, Vic 193
Symes, James 43
Symes, Mathew 43

Tack Long 104–6, 112–13
tailings 13–14 see also sludge
Talbot diggings 85
Talbot Reservoir 79
Tarrawingee, Vic
 channelising sludge in 155, 164,
  176–7
 sludge in 26–7, 136–9, 154–5,
  183–4, 188
Tasmania 42, 49, 211
Tavistock Hill, Vic 5
technology 209–10, 251–2
Telford, William 85, 96, 170
Thomson River 24

Three Mile Creek
 complaints about sludge from
  136
 conflicts along 104–6, 112
 mining at 38, 205
 pits of 36, 38
 settling dams 138, 170, 207, 230
 water supply 83–5, 120
timber 57–8
tin mining
 demographics of 41–2
 practices 48–9, 60, 132
Tooborac, Vic 9
Toora, Vic 48
tourism 234–5
Toutcher, Richard 189, 192
Trades Hall 166
Traditional Owner Settlement Act
 2010 221
Trollope, Anthony 113
Tronoh dredge 61
Trove 139
Tullaroop Creek 22, 169, 239–40
tunnels 94–5, 97, 116, 123, 125
Tuolumne County, US 41
Turnbull, Jodi 6, 38, 94, 120
Turon goldfield 76
Two Mile Creek 105

underground water 35–6, 123–4, 226
Union Gold Sluicing Company 30
United Nations 217
United Sluicing Company 170
University of Queensland 212

Vaughan, Vic 103, 171
VicMine database 6, 214
VicProd database 214
Victoria Gold Dredging Company 61
Victoria Government Gazette 127
Victorian Alps 51
Victorian Heritage Database 214
Victorian Public Records Office 127

Victorian Water Supply Department 54

Videan, James 79, 107

Violet Town, Vic 188

Wai Jung Chin 99

Walhalla, Vic 24

Walker, Arthur F. 168

Walker, William Crawford 100, 123

Wall, John 20

Wallace, John Alston
  career 97, 138
  home 245
  mining investments 96
  mining legislation and 161, 189
  water control 85, 96–8, 121

Wallace, Peter 97–8

Wangaratta, Vic
  John Bowser in 178
  sludge in 27–9, 169
  water supply 188

War on Waste 251

Waranga Basin 192

Wardle, Edward 85–6, 94, 100

Warr, Frederick 104–5

Warrandyte, Vic 95

Water Act 1989 132

water frontages 221–2

water races *see also* channelising;
  Coliban System of Waterworks;
  tunnels
    Ah Tan's 99
    Asa Farnsworth and James
      Syme's 85
    Bradfield family's 92
    Canada race 87–9
    constructing 86–7, 90–6
    crossings 120–2
    cutting races 110–12
    at Devil's Gully 94–5
    Donald Fletcher's 122
    Edward Wardle and Thomas
      Amos' 94

  effects of 102–3
  effiiciency of 70
  at Eureka Reef 67
  John Pund's 83–5, 103, 120
  John Reilly's 68–9, 86, 115
  Kangaroo–Wombat Creek race
    85–6
  LLWA 101–2
  maintaining 100–1
  McIvor Hydraulic Sluicing
    Company's 9–10
  regulations 107–8, 116–20, 122,
    125, 130–1
  at Stanley 35–9
  use of springs 122

water rights 104–35

water use *see also* dredges; puddlers;
  sluices
    drinking water 23–4, 216–17
    irrigation 192–3, 218–20
    panning and cradling 46

waterholes 178, 239

waterways *see* dams; *names of
  specific rivers and creeks*; spring
  water; water races; waterholes

waterwheels 52–7, 171

Wedderburn, Vic 71

Welshpool, Vic 48

Western Australia 229

'What is sludge?' 11

Wheeler, James Henry 159

White Hills, Vic 14, 63–4, 169

Whittle, Willie 26–7, 155

Wild Duck Creek 86

Williams, James 26

Williams, John 79, 105–7

Wimmera, Vic 97

Winter's Creek 157

Withers, William 243

Wohl, Ellen 241

Wombat Creek 85–6

women 216–17, 246

Woods, John 159

World War I 205
Wright, Clare 88, 246
Wright, Peter 40, 44, 53

Yackandandah, Vic 30, 38, 83,
    118, 169
Yackandandah Creek 68, 169
Yan Yean reservoir 70

Yankee Dam 76–7
'Yankee' party 87, 88
Yapeen, Vic 172
Yarra River 95, 188, 233
Yarrowee River 19, 144, 155, 203
Yeomans, Percy 219

Zincke, William Lawrence 160, 164

SUSAN LAWRENCE is a professor of archaeology at La Trobe University and has spent thirty years studying the goldfields. She is the author of *Dolly's Creek: An Archaeology of a Victorian Goldfields Community* and, with Peter Davies, *An Archaeology of Australia since 1788*.

PETER DAVIES is a research fellow in archaeology at La Trobe University whose work focuses on the social, industrial and environmental archaeology of colonial Australia. His previous books include *Henry's Mill: The Archaeology and History of a Forest Community* and *An Archaeology of Institutional Confinement*.